Sustaining Development in North-East India

ABOUT THE EDITOR

Anupam Hazra: [born in 1982] Presently Assistant Professor in Department of Social Work at Assam University, Silchar (Assam). Obtained Master's Degree (2005) in Social Work from Visva-Bharati University (Santiniketan), West Bengal and is currently on the verge of completion of his doctoral research on 'Rural Sanitation' from Assam University. Prior to joining academic fraternity worked with international organization like Water Aid. His areas of academic and research interest include 'Rural Development', 'Social Work Literature' and 'Development Policies'. Besides being the author of the books *India's Social Sector and Millennium Development Goals* (Rawat Publications, 2013) and *Rural India and The Emerging Developmental Challenges* (Mittal Publications, 2013), contributed around hundred papers for edited books and leading social science journals. Introduced 'Social Work Chronicle' – an international referred journal, in 2012 and also associated with editorial board of several social science journals. Presented papers in international and national seminars and is actively involved in research and consultancy projects of different development organizations.

Sustaining Development in North-East India

Emerging Issues, Challenges and Policy Measures

Edited by

ANUPAM HAZRA

CONCEPT PUBLISHING COMPANY PVT. LTD.
NEW DELHI-110059

ISBN-13: 978-91-5125-024-1

First Published 2014

Published and Printed by

Since 1974

Concept Publishing Company Pvt. Ltd.
A/15-16, Commercial Block, Mohan Garden
New Delhi-110059 (India)
T: 25351460, 25351794, *F*: 091-11-25357109
E: publishing@conceptpub.com, W: www.conceptpub.com
Editorial Office: H-13, Bali Nagar, New Delhi-110 015, India.

Cataloging in Publication Data--*Courtesy:* D.K. Agencies (P) Ltd. <docinfo@dkagencies.com>

Sustaining development in North-East India : emerging issues, challenges and policy measures / edited by Anupam Hazra.

p. cm.
Contributed articles.
Includes bibliographical references and index.
ISBN 9789151250241

1. India, Northeastern—Scheduled tribes—Government policy. 2. Economic development—India, Northeastern. 3. Women in development—India, Northeastern. 4. Education, Primary—India, Northeastern. 5. Education and state—India, Northeastern. 6. Rural development—India, Northeastern. 7. Sustainable development—India, Northeastern. I. Hazra, Anupam.

DDC 305.8009541 23

This book is dedicated

to my

Grandfather (LATE) SRI ABHILASH HAZRA

Preface

The agenda of sustainable development is complex and multidimensional. Addressing sustainable development in India is of vital importance since a large portion of the population is living below the poverty line, millions of people do not have access to basic health care and income inequalities are rising day by day and when the issue of sustainable development is discussed in the context of North-East India, it easily draws attention of the policy-makers and development experts. The North-East India comprising of eight States namely Assam, Arunachal Pradesh, Manipur, Mizoram, Tripura, Nagaland, Meghalaya and Sikkim which is recognized to be geographically isolated region, densely administered, sparsely populated, and least developed region as compared to other States of India. In the absence of major industrial establishment and other employment opportunities in the region unemployment rate, particularly urban educated youths, is not only high but also increasing rapidly. The percentage of population living below the poverty line in the region continued to remain high. Reports and statistics aptly reflect that the level of human development is significantly high in the States like Mizoram, Manipur and Nagaland and comparatively low in States of Arunachal Pradesh, Assam, Meghalaya, and Tripura. An overall scenario of the region in the post-globalization period is not very impressive as in the case of other region of the Indian union. This may be attributed to the prevailing scenario of human development in the region and attitude of the Central Government in tackling the emerging issues and challenges in the sphere of human development of the North-Eastern States. The unattended issues and problems of the past are being accumulated, multi-layered and have become multifarious. Over and above, the mounting pressures of emerging challenges of rapid transformation need to be countered. The clock is ticking fast and situation in the region is very delicate which may explode anytime from now if not tackled carefully. This edited volume, through its eighteen chapters makes a concerted effort to explore the grassroot dynamics of social sector development of North-Eastern States, with a major focus on the key social sector issues like education, health, women empowerment, etc., and explores some of the emerging issues and challenges in attaining the long cherished goal of ensuring a sustainable growth in North-Eastern region of this country.

Traditionally the process of displacement had been made otherwise acceptable to people with two justifications. The first one seemed to be partly rationale, which promised to create more employment opportunities for the displaced people. However, the second logic, that such displacement process would certainly provide an opportunity to integrate those tribal communities with the mainstream, hardly could transform their lives. Amidst lot of deprivations and continuous discriminations, there is a little ray of hope. The recently developed mechanisms and administrative processes, which include joint forest management, extension of Panchayati Raj to scheduled areas, Protection of Plant variety and Farmers' Rights Act, Biological Diversity Act, Scheduled Tribes and other Traditional Forest Dwellers (Rights over the forest) Act, may substantially protect the interests and rights of tribal communities for their own self-determination and 'Swaraj'. **Dr. Indrajit Goswami,** in the introductory chapter has highlighted the issue of land alienation and displacement of tribal communities of North-East India and its impact on their social, ethnic and cultural life with an ecological perspective.

Primary education has always been considered as one of the core component of human development. Since independence, the Government has made significant investment for promoting the access to primary education even in the remotest part of the country. But when we take into consideration the present scenario of primary education in the North-Eastern region, the progress seems to be a bit sluggish in comparison to the other parts of the country. Researches, reports and other available literature suggest that issues like high drop-outs at the primary level, lack of proper infrastructure in the primary schools, high absenteeism among school teachers, etc., are jeopardizing the efforts of the Government directed towards promoting the level of primary education in the North-Eastern States.

The slothful progress of primary education in the region seems to be putting an adverse effect on the overall progress of human development across the North-Eastern States of India. **Ms. Subhashree Sanyal** explores the current scenario of primary education across the North-Eastern States of India and puts a major focus on the emerging issues and challenges in the context of educational progress at primary level across North-Eastern States. Keeping in view the current challenges, the chapter concludes with some suggestive measures which will ensure a smooth progress of primary education in the North-Eastern region of the country.

The idea of "tribes" is particularly complex in the Indian subcontinent, where indigenous/primitive inhabitants were neither eliminated, nor quite absorbed, by the rising civilization in the course of history. It is plain enough that the demography of tribal people cannot loom large in India's overall demographic scene (the former's relative size being only a little more than 8 per cent). The 'practice' of tribal development started with the British but they never concealed their motive in following the 'tasks of mercy'. The gravity of situation was well realized at the dawn of independence. In the beginning itself, it was realized that tribals must be 'integrated' into the national mainstream—they should neither be assimilated, nor isolated. **Dr. Ajeet Jaiswal**, in his chapter discusses the current dynamics of tribal development in the North-Eastern part of our nation and has highlighted the emerging issues and constraints in the context of tribal development.

The North-Eastern region, under the shadow of conflict, has witnessed a resurgence of patriarchal values and norms, which have brought with them new restrictions on the movement of women, the dress they wear and more overtly physical violence such as rape, which is systematically used as a tactic against a particular community. All this is compounded by the long social, economic and psychological trauma of armed conflict. The chapter by **Dr. Molankal Gangabhushan and Ms. Baiakmen S. Hek** is based on the secondary source of information which attempts to bring forward the current status of women in North-Eastern region in view of the above indicators and analyse the factors giving rise to marginalisation of women.

Prof. G. Ram brings out the patterns of human deprivation in a multi-cultural society, revealing that in such society, basically, cultural groups, rather than regions and sub-regions, are deprived from human development, however, it is generally perceived that a region or sub-region by its adverse nature deprives its population from human development. It urges to focus on the cultural dynamics of people and governance, stressing that the marginalized cultural groups are historically pushed to the regions and territories which are poor in economic resources and prone to natural calamities, on the one hand, and even in the developed regions or territories such groups generally remain deprived from the resources, on the other.

The chapter by **Dr. Sumarbin Umdor** reports on implementation of Mid-Day Meal Schemes (MDMS) for primary school children in Meghalaya. Based on survey carried out in 2008, it reveals how the implementation of the scheme in the State is being affected by irregular supply of funds and foodgrains to schools and the absence of basic facility such as kitchen shed in the schools where meals can be cooked. Further, well defined norms and guidelines of the scheme such as free delivery of foodgrains to schools are not being followed, with schools having to arrange and pay for the transporting foodgrains from the supplier. An important issue raised by the survey is also with regard to meal timing in morning shift schools which comprised many of primary schools particularly in the rural areas. The survey exposes the sloppy implementation of the schemes and the failure of the State to take advantage of the programme to improve primary education in the State.

There are two-way linkages between economic under-development and insurgency. An underdeveloped material base provides little opportunities to people for gainful employment. Unemployed youths then hardly find any viable alternative for their livelihood. They perceive that the 'failed state' is the prime cause for their misery. This perception encourages them to join the non-State actors who vouch for revolutionary change. Thus economic underdevelopment helps in sharpening anti-State political discourse and strengthens the ideology of the so called revolutionary groups. Insurgent activities help in perpetuating economic underdevelopment. Militancy has created deep insecurity in the minds of the people in general and entrepreneurial class in particular. The chapter by **Dr. K. Gyanendra Singh** is an attempt to map the effect of insurgency on the economy of Manipur.

Rural Development in Mizoram was severely slowed down and affected due to two decades of insurgency (1966-86) and grouping of villages as counter insurgency strategy. The grouping of villages has its permanent repercussions where post-grouping reconstruction measures were not initiated so as to suit the people and the area of the villages that grouped. Although various rural development initiatives were introduced and implemented during post-insurgency period, its impact on rural economy is minimal. The result is that approximately all villages and the State is not economically self-sufficient. The State still depends on almost all essential commodities from outside the State. **Dr. Harendra Sinha**, through his chapter focuses mainly on the impact of village grouping as counterinsurgency strategy and the post-independence rural development programmes initiated and its impact on rural Mizoram.

A comprehensive analysis by **Dr. Anindita Sengupta** shows that in Tripura during the past five years, while number of primary level educational institutions, especially the number of primary level government schools has increased phenomenally; gross enrolment ratio of the students in the primary classes has also increased incredibly in the districts of Tripura. It also explains to the explanatory factors of drop-out of students from the primary level educational institutions in the districts of Tripura over the period from 2005-06 to 2009-10 in terms of Panel Data Regression Model and also to see the nature of relative contribution of these explanatory factors during the period of study.

The developmental experience of Nagaland has been full of challenges. Apart from its late start, geographical remoteness and inaccessibility, hilly terrain, lack of infrastructure, population composition, and scarce resource base, the State also had to face continuous insurgency, spending much of its resources on administration and related costs at the expense of development. Under this backdrop, the chapter by **Prof. Ratna Roy Sanyal and Ms. Subhashree Sanyal,** adopts a comprehensive analytical approach to track the current situation of Nagaland's sustainable growth by putting the State's current growth scenario into Millennium Development Goal framework.

Development is a multi-dimensional phenomenon. Some of its major dimensions include the level of economic growth, level of education, level of health services, degree of modernisation, status of women, level of nutrition, quality of housing, distribution of goods and services, and access to transportation and communication. In Arunachal Pradesh, the progress of socio-economic development among tribal population is not uniform. Development has both positive and negative implications on society and people. Under this backdrop, **Mr. Tame Ramya** through his chapter examines the socio-economic development processes of Nyishis, numerically largest tribe, inhabiting in five districts of Arunachal Pradesh. It tries to look at the role of factionalism and faction politics in the implementation of development interventions in Nyishi tribal village, called Hiya. It also focuses on the type of change that they have experienced with the implementation of different schemes by both Governmental and Non-Governmental agencies.

Manipur, one of the eight sisters of North-Eastern Region of India is situated in the strategic corner of the country. But, it is worthwhile to mention that information regarding the total land utilization of the entire State cannot be obtained since land records are available only for the valley and a small part of the hill areas. Due to the rapid growth population in the State, acute pressure is on the valley area where all the people belonging to different ethnic groups use to settle and concentrate with high density, thereby leading to urbanisation. On the other hand, the region accounting for 90.00 per cent of the total land area of the State is only meant for the Scheduled Tribes and is sparsely populated. Through his chapter, **Dr. Elangbam Nixon Singh** has made an attempt to analyse the growth of population in Manipur during 1951 and 2011, keeping urbanisation in view for sustainable development of the State. In fact, the chapter is purely based on various secondary data/reports published by the State government and the census data, being prepared by Census of India, from time to time.

Mr. Amitabha Ray through his chapter has put an in-depth focus on the emerging developmental issues and concerns of Jaintia Hills of Meghalaya. His chapter has not only highlighted the unnoticed beauty of Jaintia Hills but also made a comprehensive effort to explore its developmental potentiality, which in near future would be putting a significant impact on the overall sustainable growth of Meghalaya as well as on the overall progress of the North-Eastern region.

The idea of using mass media for disseminating health messages and for curing the gaps in health sector has a history of nearly six decades in the developed world. In Indian context there has been sporadic efforts but with little success. Recently in Assam, there has been a significant improvement in Assam although still poorer than national average but a significant change nonetheless observed. On scrutiny, it has been found that an assiduous intervention of conventional media has been able to reach out to the people to a significant extent despite the stumbling blocks of illiteracy and incapability to access conventional media on the part of rural poor. It has happened largely due to the intensive approach adopted by the development planners assuming that it is their responsibility to reach the rural as well as informational poor. The chapter by **Dr. Silajit Guha and Mr. Biraj Kanti Shome** shows that this intensive approach has taken care of the basic needs, but it goes without saying that additional efforts are required to push the borderline even further; the sector now badly requires a grass root level communication approach taking the culture specificity and complexities of the region into mindscape. Unless the development planners start designing the messages with the dictum in the mind that communication is nothing but the social interaction of the messages, there is a little chance that the intervention of the mass media would be able to remedy the pathetic situation that exists in the health sector of Assam as well as the region.

The sexual and reproductive health of pregnant woman living with HIV/AIDS is fundamental to their well-being, Gender plays an important role in determining a woman's vulnerability to HIV infection and her ability to access treatment, care and support. All women have the same rights concerning their production and sexuality but women living with HIV/AIDS require additional care and counselling during their reproductive life. **Dr. Grace Hmar's** chapter addresses the specific sexual and reproductive health needs of women living with HIV/AIDS and contains recommendations for counselling, care and other interventions. Where sexual and reproductive health services tailored to the needs and circumstances of people living with HIV are in place, not only do the lives of the people with HIV stand to benefit but Global HIV prevention efforts will benefit as well.

The contribution of women in running a house has been considered as the most important role of women in a traditional society. The effect of work and contribution of women in making a home is intrinsically related to the development of the entire economy of a particular society. Often this fact is overlooked and women's work both in the household and outside remains unaccredited. This kind of work is also many a time unpaid. However, this contribution of women's work and its significance is clearly evident in case of the tribal and matrilineal societies. **Dr. Jyoti Prasad Saikia and Ms. Anannya Gogoi** through their chapter provides a description of the nature and consequences of the role played by Karbi women in development of their society in the Karbi Anglong district of Assam of the North-Eastern part of India. An attempt has also been made to observe the indigenous knowledge of the Karbi people which is visible in agriculture and its allied activities.

The study-based reflections by **Dr. P.C. Sikligar** focuses on the various aspects of handloom cottage industry in terms of problems, importance of special project (under SGSY: Swarna Jayanti Grameen Swarozgar Yojana), and strategy for further development of handloom cottage industry in Manipur. This study was carried out in Imphal districts both east and west to understand the problems and scope of the handloom cottage industry in the State where women are working under the shadow of self-help groups which also covers focus group discussion with the members of different 20 self help groups and are engaged in weaving.

In the last chapter, **Mr. Anupam Hazra** addresses one of the emerging crucial developmental agenda, which is gaining global attention in the context of sustainable development, i.e., the issue of sanitation. In the beginning, his chapter makes an attempt to explore the importance of sanitation in the context of attainment of Millennium Development Goals. Later the chapter presents the country's overall sanitation situation and highlights the current dynamic scenario of rural sanitation and the emerging issues and challenges that Assam is facing in the sphere of promoting rural sanitation across the State. The chapter concludes with some suggestive measures, which the author thinks, would be instrumental in promoting sanitation across the rural parts of the State.

Anupam Hazra

Acknowledgements

I would like to express my deepest sense of gratitude to my parents, Smt. Manju Hazra and Shri Debnath Hazra and to my sweet 'Didibhai', Smt. Annapurna Mondal for their blessings. I duly acknowledge the support and encouragement provided by my loving brother, Mr. Nirupam Hazra, I am grateful to all my critics who have always inspired me to put more sincere and intense efforts towards my every academic endeavour with their 'valuable' criticism.

Anupam Hazra

A Message for the Readers

Partition of Modern India: A View from 'North-East Lens'

Nirupam Hazra

There is something ominously unsettling about India's 65th Independence Day, especially when taken into account the events preceded and followed it. It creates a sense of *déjà vu.* When India got independence in 1947, it was preceded and followed by riots, bloodshed and displacement. It may sound hyperbolic or disproportionately exaggerated. Human tragedy like Partition has no historical parallel in independent India. Partition resulted in death of thousands of people and displacement of millions. Riots followed by more riots, death followed by more deaths. The fanaticism and collective madness of Partition are poignantly portrayed in the works of contemporary writers. In the works of Manto we find vivid description of the horror of partition, beyond the mere drab official statistics of death and displacement. There is a host of other literary narratives of the event of partition which comes under the rubric of 'partition literature'.

It was a story of infant India. It was a story of that India which was yet to become world's largest secular democracy, which was yet to get tag of 'next superpower.' Then it had many villains to be blamed for the disaster.

But now, let us focus on the events that surround India's 65th Independence Day. Communal violence broke out in Assam in early July. Gradually, it took the shape of one of the worst man-made disaster of modern India. Hundreds (officially not even hundred) of people got killed, four lakhs are displaced. Perhaps, first time in independent India, four lakhs people are living in relief camp because of a communal strife. This communal violence is followed by another episode of violence. Though this time the violence is not as explicit as what took place in Assam. But it is more shameful, more outrageous and most unfortunate. No sooner the physical violence subsided, the ripple effect of it was felt in other parts of the country. It led to the exodus of North-Eastern people from so-called cosmopolitan cities like Pune, Bangalore and Vadodra. Thousands of North-Easterns are running for their life, leaving behind livelihood, jobs and career. The image of thousands of panic-stricken people, precariously hanging in overcrowded trains, inevitably brings back the image of Partition. That time, it was formation of a new nation or a colonial conspiracy or a rightful demand. But this time it is a fragmentation of a nation, not literal, but psychological, not political, but emotional, not geographical, but demographical. And what makes these events more shameful, is the fact it is happening in world's largest democracy. It is happening in a country where guests are compared with God, but fellow-citizens are confused with foreigners. Such events put a question mark on the pluralistic credential of our country. It threatens the very notion of *swikriti* that Dr. Amartya Sen talked about in his *The Argumentative Indian*. A nuanced and critical examination of the events that are taking place in our cosmopolitan cities exposes the shameful attempt of cultural homogenization. It is true that as soon as things start to get 'normal' all of us will forget. There won't be any grand narrative of this epic suffering which eventually may bring a moment of national catharsis. In the age of 140 character narrative it will soon fade into oblivion. It will find an eternal place only in the psyche of those who were bruised and brutalized.

Had such events taken place in a Muslim country under a dictator, it would have been termed as a humanitarian crisis by UN. Or in a worst scenario, America would have intervened (read invaded) to prevent such 'racial persecution'. But, remember India is world's largest democracy and emerging superpower. And don't forget that in world's largest democracy state can indiscriminately kill the tribal, army can rape and murder the women and police can routinely eliminate the citizens with impunity. It is Incredible INDEPENDENT India!

Contents

List of Contributors

Gogoi, Anannya is Assistant Professor in Department of Sociology at Dibrugarh University, Dibrugarh-786004; Assam.
Email: anannyagogoi19@gmail.com

Goswami, Indrajit is Director in Department of Management Studies at INFO Institute of Engineering, Kovilpalayam; Coimbatore-641107.
Email: idrgoswami@gmail.com

Guha, Silajit is Associate Professor in Department of Mass Communication at Assam University, Silchar-788011; Assam.
Email: silajitguha@gmail.com.

Hazra, Anupam is Assistant Professor in Department of Social Work at Assam (Central) University; Silchar-788011; Assam.
E-mail: anupam688@yahoo.co.in

Hazra, Nirupam is Post-Graduate Scholar in Department of Social Work at University of Delhi. Email: hazra.nirupam@gmail.com

Hek, Baiakmen S. is a Research Scholar in Department of Social Work at Assam University, Silchar-788011, Assam.
Email: shashahek@gmail.com

Hmar, Grace is Assistant Professor in Department of Social Work at Indira Gandhi National Tribal University, Manipur (Regional Campus).
E-mail: Grace_laltlinzo@rediffmail.com

Jaiswal, Ajeet is Assistant Professor in Department of Anthropology at Pondicherry University, Puducherry-605014; India.
Email: rpgajeet@gmail.com

Molankal, Gangabhushan is Assistant Professor in Department of Social Work at Assam University, Silchar-788011; Assam.
Email: gangabhushan@gmail.com

Nixon Singh, Elangbam is Associate Professor in Department of Management at Mizoram University, Aizawl-796009; Mizoram. Email: singhnixon@yahoo.co.in

Ram, G. is Professor in Department of Sociology at Assam University, Silchar-788011 (Assam).
E-mail: gramsoc@gmail.com

Ramya, Tame is a Research Scholar in Arunachal Institute of Tribal Studies (AITS) at Rajiv Gandhi University [A Central University], Doimukh-791112, Arunachal Pradesh.
E-mail: tameramya@rediffmail.com

Ray, Amitabha is a Deputy Adviser in Planning Commission, Government of India; New Delhi-110001. Email: araypc@gmail.com.

Roy, Sanyal, Ratna is Retired Professor from Department of History, under University of North Bengal, West Bengal-734013.
Email: rroysanyal@yahoo.co.in.

Saikia, Jyoti Prasad is Associate Professor in Department of Sociology at Dibrugarh University, Dibrugarh-786004; Assam.
Email: saikia_jyotiprasad@yahoo.co.in

Sanyal, Subhashree is Assistant Professor in Department of Social Work at Visva Bharati; Sriniketan-731236.
Email: subhashreesanyal@gmail.com

Sengupta , Anindita is Assistant Professor in Department of Economics at Hooghly Women's College, Hooghly-712103; West Bengal.
E-mail: a_s_gupta2@rediffmail.com

Shome, Biraj Kanti is a Research Scholar in Department of Mass Communication at Assam University, Silchar-788011; Assam.

Sikligar, P.C. is Associate Professor (Research and Training Division) at National Institute of Rural Development, Rajendranagar-500030; Hyderabad. E-mail: pcsikligar@yahoo.co.in

Singh, K. Gyanendra is Assistant Professor in Department of Humanities & Social Sciences at National Institute of Technology, Aizawl-796012; Mizoram. E-mail: gyanendra_eco@yahoo.com

Sinha, Harendra is Associate Professor in Department of Political Science at Government J. Buana College, Lunglei; Mizoram-796701.
Email: dawlasinha@gmail.com

Umdor, Sumarbin is Associate Professor in Department of Economics at North Eastern Hill University [NEHU], Shillong-793022; Meghalaya.
Email: sumdor@nehu.ac.in

1

Development and Displacement: The Eco-system of Tribal Dwellings in North-East India

Indrajit Goswami

Indigenous technology, indigenous methods and indigenous knowledge might have evolved as popular jargons to familiarize the contemporary approaches of sustainable development. But 'inclusive growth and development of indigenous populations'—an implied and inseparable principle of sustainable development process holds almost little importance in India. The hidden agenda, behind argument of 'mainstreaming of those communities' had been the expropriation of their resources in the name of their development. It is an undisputed truth that where the nature is rich, the local people are still living below poverty line. Since 1980, almost 12-13 per cent of the total forest area in the entire country, including North-East regions, diverted for mining and related industrial activities. The people, mostly the indigenous tribal populations, were relocated and displaced without adequate compensation and rehabilitation. That consistent and sustainable exploitation might have been one of the factors responsible for the social and political destabilizations and ethnic violence in several parts of North-East regions.

Traditionally the process of displacement had been made otherwise acceptable to people with two justifications. The first one seemed to be partly rationale, which promised to create more employment opportunities for the displaced people. However, the second logic, that such displacement process would certainly provide an opportunity to integrate those tribal communities with the mainstream, hardly could transform their lives.

Amidst lot of deprivations and continuous discriminations, there is a little ray of hope. The recently developed mechanisms and administrative processes, which include joint forest management, extension of *Panchayati Raj* to scheduled areas, Protection of Plant variety and Farmers' Rights Act, Biological Diversity Act, Scheduled Tribes and other Traditional Forest Dwellers (Rights over the forest) Act, may substantially protect the interests and rights of tribal communities for their own self-determination and *Swaraj*.

North-East India is predominantly inhabited by many indigenous tribes who have been inheriting their ancient customs and traditions. Those tribes have had a direct association with their natural surroundings. Traditionally they have been living by sustaining the health and bio-diversity of their own local ecosystems. It is they who nurture and guard these ecosystems most effectively. Right from their values and customs, lifestyle and food habits are greatly defined and shaped by their surrounding environment, especially by forest, land and water resources. Each State is the home of a number of tribes, who have their own distinctive culture, language and traditions. There are more than hundred types of tribal communities in the seven North-East (N-E) States. Some of the prominent tribal communities of the region include Nagas, Khasis, Jaintias, Mizos, Boros, and Cacharis. A non-exhaustive list of distribution of tribes in seven North-Eastern States is given below:

India State	*Dominant Tribal Communities in the particular State*
Arunachal Pradesh	Aka, Apatani, Deori, Khamti, Khamba, Memba, Mishing, Miri, Khowa, Mishmi, Lisu, Miji, Monpa, Sulung, Tangsa, etc.
Assam	Boro, Kachari, Dimasa, Karbi, Mising, Koch, Rajbonshi, Garo, Gangte, Hajong, Khansi-pnars, Paite, etc.
Manipur	Himol, Anal, Angami, Chisu, Chota, Gangte, Kabui, Khoirao, Kom, Lumjong, Monsang, Maram, Marim, Mao, Mayong, Paite, Pusum, Ralte, Sema, Simti, Sukte, Thangkul, Thadao, Vaiphei, Zou, etc.
Meghalaya	Khasi-jayantia, Garo, Karbi, Lalung, Hajong, Biate, Koch, etc.
Mizoram	Lushai, Hmar, Pawi,Paite,Riwang, etc.
Nagaland	Ao, Sema, Konyak, Lotha, Angami, Chang, Rengma, Phom, Pouchari, Chakesang, Zeliangrong, etc.
Tripura	Chakma, Tripuri, Reang, Jamatia, Lushai, Kuki, etc.

Once, the British described the Indian sub-continent as an 'ocean of trees'. People of India, especially the indigenous tribal communities had played a vital role in nurturing these tree wealth, and Dietrich Brandis, specially brought in from Germany as India's first Inspector General of Forests admired the contribution and insisted that a substantial part of forest land be left in charge of people as 'village forests' when the forest laws were formulated. However, interference by successive governments, since British period, in the customary rights of the tribals over their land and minor forest produce; economic, industrial and urban development programmes and unchecked entry of non-tribal traders and money lenders into socio-economic life of the tribals, have all contributed a great deal in gradually weaning away the tribals from their lands. This had the unfortunate consequences of destroying people's motivation in safeguarding forests and other natural resources. At Mahatma Gandhi's insistence, the Congress had promised to change the anti-people forest policy after independence. But this did not happen. The dominant economic and political elite in the country are aggressively pursuing policies to promote industrialization and commoditization of basic services and natural resources like land, water and forests. The State institutions, structures and policies are being drastically altered, often beyond recognition, to provide a conducive environment for the large scale displacement of indigenous people (Invally, 2008).

Traditional Land Holding and Management

Immediately after independence, the ruling politicians had promised people greater access to forest land and resources. But the forest policy (1952) asserted the primacy of 'national' interests and stated that people were to have no special rights just because of the fact that they were born and had been living in the vicinity of the forests. The policy further stated that it would be improper to concede any rights on the forest lands or forest produce to the people. That had been used all along to exploit the people, either to force them into bribing the officials or transfer their natural possessions to the others. After 36 years, the new forest policy of 1988 recognized the importance and role of forest dwellers, mostly the tribals, and emphasized the need to protect the remaining natural forest lands. Later, the Forest Rights Act 2006 was adopted in December 2006, but it hardly could bring qualitative changes in the existing practices. The Relief and

Rehabilitation Policy had been revised twice since 2003; the latest one was made public in November 2007. But those efforts neither could address the customary laws of tribes in N-E, nor could integrate them.

It is relevant to mention here that the ownership pattern and land management system among different tribal communities are markedly different from each other. While the Angami of Nagaland (D'Souza, 2001) and the Dimasa of Assam (Bordoloi, 1986) had a combined tradition of individual and community ownership, others like the Aka of Arunachal Pradesh had no customary practice of individual ownership (Fernandes and Pereira, 2005). However, very interestingly there was a common tradition among all tribes, where the community had the authority even to govern individual ownership. During the post-independence period, the imposition of individual-based land laws brought into the practice of privatization, which facilitated the land alienation process. As viewed by Shimray (2009), the attempts of governments in the States of Manipur, Tripura and Assam to change the laws in favour of individual ownership, have not only facilitated transfer of tribal land to non-tribals, but also lead the way of tribal exploitation in those States. According to D'Souza (1999) land is basic to the ethnic and traditional tribal identity as well as to their modern economy because more than 70 per cent of the people depend on it for their sustenance. The immigrants, mostly from Bangladesh, Bihar, Jharkhand and Nepal had been in low paid jobs such as on construction sites, as rickshaw pullers and others, occupied the common property resources (CPR) that were tribal livelihood and the centre of their economy, culture and identity. Most of those immigrants were landless agricultural labourers before their arrival in the North-East. They used to live in a feudal system of lack of land reforms, low wages and poverty. That was the push factor and the fertile land in the Brahmaputra Valley appeared as a pull factor (Majumdar, 1987). The push and pull factor dynamics was further fuelled by the new legal system in the post-independent India, which made encroachment easy and consequently most of the immigrants occupied CPRs belonged to the tribes. The conflict between community-based customary laws and State land laws was never resolved and thus recognized only individual ownership and treated land without *pattas* as State property. Such conflicts and ambiguity made it easy for the immigrants to encroach on tribal and other CPRs. Further, the process of privatization had a smooth passage through the mechanisms of indebtedness, sale, mortgage, sub-tenancy, exchange, displacement and *benami* transactions, which continued unabated. There are certain constitutions, legislative and executive measures for the prevention of such process but because of lack of awareness among tribal people and political unwillingness the situation never changed nor improved. Most importantly, since most of such transactions take place through informal agreements, there are no accurate and complete records to safeguard the alienated and displaced tribals. It has been observed that the ambiguity in the existing legal systems has promoted the privatization process mostly in all N-E States. These laws not only facilitated the alienation of CPR to non-tribals but also made it easy for the State to transfer such CPR to development projects. The similar instances have been found in Assam, Arunachal Pradesh, Tripura, Manipur, etc. It has been observed that the Tripura Land Reforms and Land Revenue Act 1960 (TLR&LR) recognized only individual ownership. As a result, most of the community owned lands were not recognized and registered (Bhowmik, 2003). The CPR system had a more equitable and gender-friendly distribution of labour and resources. Over the period of time, the State run machineries, instead of protecting and safeguarding such systems, widened the gender gap by recognizing the men of tribal households as head of families. In Assam and Meghalaya, the Coffee and Tea Board offered loan and subsidies to individuals alone. Unfortunately, the men, especially from elite groups from within the tribal communities, enjoyed the benefits. They manipulated the system and transferred CPR into their names (Fernandez and Bharali, 2002).

Traditionally tribals have developed their own system of land classification and clearly demarcated community and private lands. Such lands were easily identifiable. Such practices were followed and preserved for generations. But North-Eastern Council (NEC) on land use classification ignored the age-old system and imposed another classification based on its own convenience and political manipulation. NEC brought into the terms in practice, such as 'area of non-agricultural uses', 'barren and uncultivable land', 'permanent pastures and other grazing land', 'cultivable waste land' and 'others' (Nongkynrih, 2008). The Constitution of India has provided special safeguards to the customary beliefs and practices of tribes of N-E. But the States and political systems have failed to protect the interests of tribes and the traditional classification of lands. The gap between Constitutional safeguards and current practices have sustained the discrimination, alienation and displacement of tribes from mainstream of life and otherwise derecognized the model of beneficial community living for a better balance between development and eco-system.

Over-exploitation of NTFPs

The loss of forest lands translates directly to losses in livelihoods, especially of tribes, who depend largely on forest regions for their day-to-day survival. It was a time when more than 50 per cent of their household incomes had been from off-farm activities. Commercialization of forest fauna and flora was not a new phenomenon to tribal economics, but it was followed and maintained through a customary system, where maintaining equilibrium in nature was the part of tribal way of life or culture. That commercialization culture of tribal helped to develop a link between forest products, poverty reduction and sustainable forests. Moreover, sustainable forests naturally regenerated resources in order to improve the quality of life of tribes as well as it protected and nurtured the biodiversity of the regions. Over the years, due to unabated land alienation and displacement, the tribal households tend to invest very little effort in cultivation and rely on non-timber forest products (NTFPs). In the mean time, due to price rise, the commoditization of NTFPs was initiated by traders, mostly non-tribals and outsiders. As a result the local tribes and resources were overexploited. Due to this overexploitation, the NTFPs species had been facing greater deficit because regeneration is affected by collection of seeds and fruits from existing forest species. Lack of government support, declining ratio of community land holding and sole dependency on NTFPs might have otherwise endangered the ecology and environment of the region. The long-term sustainability of forests is no longer a preferred way of life to the tribes, because they have been induced to survive on short-term benefits. Pachauri (1990) has rightly observed that survival of forests would be impossible under State ownership and the conditions of forests would be constantly deteriorating unless the forest dwellers, the tribals, are given management and conservation rights.

The *Jhum* Cycle

Agriculture has been the main livelihood among the hill and plain tribes. In addition to paddy cultivation, *jhum* (shifting cultivation) is carried out by many tribal groups. *Jhuming* is one of the most ancient systems of farming and it is uniquely linked with the ethos and the social and cultural values of the tribal communities (Borthakur, 1992). There has been a general understanding among policy makers and planners that *jhuming* have adverse effects on ecology and conservation, including destruction of soil fertility, soil erosion in the upper catchments resulting in sedimentation of waterbodies, destruction of wildlife and natural habitat, and flooding. They believed that *jhum* cultivation was bad and thus should be eradicated. However, the fact is that the *jhum* cultivation cycle is favourable for viable land use and for conservation of natural resources. Currently the *jhum* cultivation has been grossly affected and reduced due to alienation of land dwellers and expansion of urban areas and commercial encroachments. The restoration of *jhum* cycles had been difficult because of unresolved conflict and dilemma among the policy makers and planners. While a group of them recognizes the practice as 'agriculture in forest land', the other group holds the views of the territory as 'forests on agricultural land'.

Ecological Imperatives

Encroachment onto forestland is a serious threat to forests and their conservation. The situation is alarming in some States and needs strong political will to address the problem, which has caused loss of forest area and created a tool for perpetual degradation of forest resources (Kumar, 2002). In addition, the adopted model of conservation has shattered the traditional harmonious relationship among tribals, nature and wildlife. In the name of conservation of wildlife, it has barred participation of forest communities in forest management. The adopted models have forcibly reduced the dependency of forest dwellers on forest resources by denying them access to forests and breaking the chain of inter-dependence that existed for centuries. At the same time, the concept of extracting profit from forest resources opened up the forests to other forces of destruction. The displacement of forest dwellers thus had a remarkable impact on the biodiversity and wildlife conservation programmes.

High biological diversity is often related to the forest cover of a region. Most of the North-Eastern States have more than 60 per cent of their geographic area under forest cover, a minimum suggested coverage for the hill States in the country. North-East India has 64 per cent of its total geographic area under forest cover and it is rapidly disappearing. There was a decrease of about 1,800 square kilometres in the forest cover between 1991 and 1999 (FSI, 2001). More worrisome is the fact that the quality of the forest is also deteriorating, with the dense forests (canopy closure of 40 per cent or more) becoming degraded into open forest or scrub. Though there is a succession of several edaphic formations, a vast area of land has already been transformed into barren and unproductive wasteland.

Issues and Challenges

The land alienation and displacement is not a new phenomenon. The process had been in practice even before British initiated the process of land grabbing from communities. That might be one of the important reasons, which later manifested in the conflicts and anti-colonial struggle led by indigenous tribal communities during period of East India Company's rule. Example of some major tribal movements in North-East India for land and forest rights during British period, includes, (i) *Khasi* rebellion (1829-33) and (ii) revolt by *Kacha* Nagas of Cachar district, Assam, led by Shambhudan (1882). Unfortunately, we have denied to take lessons from those movements and consistently refused to acknowledge the natural rights of tribal and other indigenous people on the forest and other cultivable lands. British failed to resolve the issue and had worsened the conditions by institutionalizing The Forest Acts of 1878 and 1927. Even after independence, we have failed to come out completely from the obsession of British legacy. Neither we could adequately check the alienation, displacement and deprivations of those people, nor could we create an integrity and synergy among political, administrative and legal systems towards ensuring their traditional wisdom and choice of living. In the name of social justice, empowerment and sustainable development, we have only widened the gap between tribal and other indigenous communities and so-called civilized and urbanized societies.

During the post-British era, the philosophy and thoughts of Mahatma Gandhi and Jawaharlal Nehru had been believed to have led the development of the nation and its polity. However, in reality Gandhi's indigenous model of building a nation of village republics and making a self-reliant society by recognizing the uniqueness and wisdom of indigenous people were not taken seriously. Instead we followed the West and imported all possible models and modes of industrialization and urbanization. In the pursuit of democracy and socialistic republic, we have only achieved social and economic inequality and disparity. Behind the development veil, we have otherwise facilitated alienating of tribal and other indigenous people from the lands and natural resources.

The first Prime Minister, late Jawaharlal Nehru laid certain fundamental principles for the welfare and development of tribal communities (*Orissa Review*, February-March, 2005). Those principles have been as follows:

1. Tribal people should develop along the lines of their own genus and we should avoid imposing anything on them. We should try to encourage in every way their traditional way of living;
2. Tribal people's rights in land and forest should be respected.
3. We should try to train and build up a team of their own people to do the work of administration and development. Some technical personnel from outside will no doubt be needed especially in the beginning. But we should avoid introducing too many outsiders into tribal territory; and
4. We should not over administer these areas or overwhelm them with a multiplicity of schemes. We should rather work through and not in rivalry to their social and cultural institutions.

The Constitution of India has also made definite provisions for the welfare and safeguard of tribal people throughout the country. Articles 15 (4) 46, 244 (1) and 339 of the Constitution of India speak of special provisions meant for the administration and control of scheduled areas and tribals therein, for their welfare and protection. The policy of respective State governments towards scheduled tribes is governed by the broad directives laid down in the Constitution of India. It expects that "The State shall promote with special care the educational and economic interest of the people and in particular of scheduled castes and scheduled tribes and shall protect them from social injustices and all forms of exploitation".

Later, we have got several commissions, committees, policies, legislations and mechanisms to address the issues and problems related to tribal communities, their life, existence and development. Except adequately in first years' plan, all other five years' plans have kept financial provisions and schemes for the welfare and development of those sections of society. However, we have miserably failed in ensuring their rights and bringing in a robust system and mechanism for their economic and social development as envisioned by Mahatma Gandhi and Jawaharlal Nehru. It has been very clear that we have repeatedly failed to understand the actual issues and problems of those people. In political, social, cultural and administrative forums, we (the externals to their communities) have identified illiteracy, health, poverty and unemployment as four major problems of tribal and indigenous people. Instead of understanding the perennial issues of those communities, our policy makers and planners have always tried to find out a budget-linked problem-solving or development models. Before we could actually understand our limitations and deficiencies, we surrendered to globalization and its compulsive and intoxicated drive towards cultural diffusions. Tribals in India have had their own significant degree of cultural and ethnic diversity. Their age old existence, way of life and futuristic ambitions has always been well-defined and well-integrated with the naturally viable and sustainable modes practices and customs. Their very basic socio-cultural, economic and ethnic existence had been well balanced with the surrounding ecosystem and biodiversity reserves. Such a system existed hundreds of years because 'community living' and 'community landholding' were two fundamental reasons. The new economic regime has led to privatization and marketization of economy and thus it has been treated as powerful threat not only to the survival of tribal communities but also to the conservation and preservation of forests, natural resources and biodiversity. One of the main issues, which has been evolved and intensified inherently, might be 'identity crisis' of the tribal and indigenous people in North-Eastern regions, not merely their 'poverty' and 'unemployment' issues *per se*. Such an inherent powerful force, may be referred as 'ethno-nationalism', might have led people to raise new claims to territory and political control over local governance and resources. Such issues may require some thoughtful insights and conscientious efforts of concerned stakeholders, rather than some budget-oriented developmental schemes and ornamental opportunity for their participation in planning and execution of development processes.

Immediately after independence, land reforms were considered as an important prerequisite for nation building. Accordingly, India's initial Five-Year Plans allocated substantial budgetary amounts for the implementation of land reforms. It could substantially abolish large landholdings in several States but it could not curb the encroachment of community lands by individuals in tribal dominated areas. Even after sixty-four years of achieving political independence, we couldn't safeguard and protect the traditional and natural rights of tribal people on their land and we couldn't ensure their economic independence by giving them rightful access to NTFRs.

In order to establish peace, harmony, integrity and socio-economic development in the North-Eastern regions in the country in general and for restoration and recognition of tribal way of life and culture, we have following few challenges to be accepted and acted on.

Challenge-1: Self-determination of tribal and indigenous people (TIP)

The extension of provisions of Panchayat Raj to Scheduled V areas has conferred full rights on local communities over minor forest produce (MFP). However, two MFPs, such as *bamboo* and *tendu* were excluded from their accessibility rights. It was unjustified. First tribal self-rule and their full right over MFPs to be restored and protected and then as envisaged, the Joint Forest Management programmes would be effective.

Challenge-2: Legal sanction to community landholdings by tribal and indigenous communities, their mutual co-existence with forest areas, and their economic and livelihood dependence on MFPs

The restoration and protection of CPRs is very essential to make such communities self-reliant and to facilitate them to live as per their own wisdom and interests. Along with such restoration efforts, the economic freedom is to be given to TIP for harvesting, collection and selling of MFPs without any intervention of middlemen or brokers.

Challenge-3: Prohibition of transfer of tribal and forest lands and its conversion into commercial and industrial areas

The key question, of course, is how to regulate encroachment on forests. The rights of forest have not been properly recorded. The land under cultivation for many generations had not been assigned to them. Such prohibition has never been possible because of influential nexus between political elites and administrative mechanisms. In case such transfer or conversion is unavoidable and very essential for larger social and economic interests, the TIP must be properly rehabilitated and compensated.

Challenge-4: Preservation, promotion and development of indigenous knowledge of TIP in agriculture, cultivation of forest produce, art and culture, and entrepreneurial endeavours

The Scheduled Tribes and Other Traditional Forest Dwellers (Rights over the Forests) Act of 2006 (TFRA) has ensured TIP, along with other rights, their full right of access to biodiversity and community right to intellectual property and traditional knowledge related to biodiversity and cultural diversity. However this right has not been fully exercised by TIP because of inadequate coordination between TFRA and the Biological Diversity Act (BDA). Instead of having separate 'Biodiversity Management Committees' (BMCs) under BDA, the *Gram Sabhas* under TFRA should have been given the assigned functions of BMCs.

Challenge-5: Maintaining cultural diversity, integrity and mutual co-existence of TIP with other communities

Instead of solving the 'identity crisis' and 'ethno-nationalism' issues, the political parties might have used those for narrow political gain. As a consequence we have witnessed ethnic conflicts in N-E regions. It has become a perennial problem and threat to national security and integrity of the region.

The following few action implications are suggested for stabilizing the issues, for meeting the challenges and for adequately solving the existential as well as developmental problems of TIP.

Action Implications

Restoration and Management of Tribal Land

Considering the absence of accurate record of land under possession and use of tribals, as also several cases of succession, new allotment of land, acquisition and transfer of land, most of which remained non-used unrecorded. A comprehensive survey needs to be undertaken to record accurate possession of lands by tribes. Tribals need to be properly educated about the importance of proper record of their lands and registration of land transfer taking place as a result of sale, mortgage, partition, etc. This will help in preventing their exploitation and land transfer, therefore need to be compulsorily recorded within a specified time frame and the parties acquiring possession of tribal land should be made accountable for the same. Also the land records may be updated to have an accurate data regarding actual possession and transfer of tribal lands. The local level committee comprising of president of the Gram Panchayat, other relevant government officials, representation of NGOs or social worker be constituted and made accountable to detect cases of land transfers and reporting thereof to the competent authority.

All the land transfers made in favour of non-tribals may be derecognized and land may be restored to the concerned tribals pending the legal course. Restrictions may be imposed on all suits against restoration of possession of lands to the tribals by the courts criminal proceedings may be initiated in all cases of illegal transfer of lands from tribal to non-tribals. Formal registration may be made compulsory in all cases of land transfers permitted under the existing rules. A general prohibition may be imposed on all transfers of tribal lands by way of sale, mortgage, gift, lease, subletting or exchange, unless otherwise permitted by the district collector. Provision of free legal aid may be made with a view to restore tribals the possession of their lands. In order to save the tribals from money lenders, alternative sources of credit may be made available and for this purpose network of cooperative societies and rural banks may be established in all tribal areas.

Considering the uneconomic holdings, concentrated efforts need to be made to enhance the productivity of tribal lands through the provision of production credit and supply inputs. Numerous welfare and development programmes have been initiated for the benefit of tribals but about which they have neither knowledge nor access. The information about these needed to be disseminated and tribes may be helped to participate in such endeavours. The officials need to be more serious about cases of land alienation submitted for consideration of restoration of land. The dynamic and enterprising tribals may be identified and encouraged to participate actively in the efficient utilization of forest and land resources and diversification of tribal economy. The dislocation and displacement of tribals due to infrastructure, industrial and development projects, need to be compensated by way of rehabilitation subsidy, development premium or share in the returns of the projects located in tribal areas till such tribes become more or less self-sufficient.

Restoration of Community Forest Land and Resources

The implied challenges in the Tribal Forest Right Act 2006 (TFRA) may be turned into empowering opportunities for forest dwelling communities, especially tribes, by engaging them in conservation of biodiversity and thus protecting the forest eco-systems. The implementation and execution of provisions in the Act requires a robust management system, which would involve people living close to nature, through the agency of *Gram Sabhas* (village committee). The *Gram Panchayats* (village councils) must constitute a village level Biodiversity Management Committee (BMC) (under jurisdiction of the Biological Diversity Act 2002) and must work in collaboration for the cause of environment, livelihood and eco-friendly community living. Since *Gram Sabhas* are fully empowered to execute the plan of works under the National Rural Employment Guarantee Scheme, it can effectively link the NREGA plans to Village Biodiversity Committees. It requires a political willingness to ensure rights granted under TFRA, where local communities would have right over lands and access to all forest lands, including reserved forests, protected forests and protected areas such as sanctuaries and national parks. If TFRA is implemented properly, it can effectively check further deforestation or takeover of forest land by outsiders.

The above aspirations are neither unachievable nor impossible. However the current disparity in administrative mechanisms (Table 1.1) could have been the major obstacle in achieving the desired aspirations. What could be the justification behind different patterns of administration for tribes in North-East? Empirically, politically or otherwise there might be different reasons but it had been transparent that the very basic biological and anthropological attributes of tribes were never given due importance. Hardly there is any comparative study, which can throw light on comparative advantages and disadvantages of different pattern of administrations. In Assam the threat of loss of tribal land is greater than any other tribal areas

Table 1.1: Types of Administrative Structures for the Tribes of N-E India

State	*Special Constitutional Provisions*	*Administrative Structure*
Arunachal Pradesh	Art. 371H	No Autonomous Council. The State has adopted Panchayati Raj
Assam	Sixth Schedule read with Art.371H (for scheduled areas only)	Three Autonomous Councils: (i) Karbi-Anglong (ii) North Cachar Hills (iii) Bodo Territorial Council
Manipur	Art .371C	The Manipur (Village Authorities in Hill Areas) Act 1956
Meghalaya	Sixth Schedule	Three Autonomous Councils: (i) Khasi Hills (ii) Jayantia Hills (iii) Garo Hills
Mizoram	Sixth Schedule to read with Art. 371G	Three Autonomous Councils of Pawi, Lakher and Chakma, and other areas without autonomous councils.
Nagaland	Art. 371A	No Autonomous district councils
Tripura	Sixth Schedule	Tripura Tribal Area Autonomous District Council, Khumulwang.

Source: Nongkynrih (2008).

(Nongkynrih, 2008). Despite the existence of three autonomous councils, the situation had yet not been improved. Hence, neither autonomy nor any special constitutional provision or safeguard would be effective in fulfilling our aspirations. Rather it requires mass empowerment and strong CPR protection mechanisms. Tribal self-determination must be respected and to be incorporated in the administrative machineries as a guiding principle for determining tribal affairs and development. The case of Ri Bhoi district in Khasi hills must be followed as a model to be replicated or adapted in different tribal areas in North-East (Jamir and Nongkynrih, 2002:95). This may be found effective to counter timber mafias, wildlife poachers and other exploiters of forest resources. If so happened, displacement of tribes and encroachment on forest and tribal lands will be prohibited. Where land under cultivation for many generations should be returned to local communities and should be preserved from sale or transfer of any kind. It is possible to create a synergy between legal, political, social and administrative mechanisms for protecting the interests of indigenous communities and to ensure their own 'self-determination' and *Swaraj*. This synergy is very much needed and solicited for the survival of indigenous people and for restoration and protection of eco-system and biodiversity in the North-Eastern regions.

Conclusion

Throughout North-East India, encroachment of forest and community lands by migrants, local elites and governments have resulted large scale alienation and displacement of indigenous communities. Several promises have been made and bunch of policies and legislations have been brought into existence, but there is hardly any remarkable improvement in restoring local cultivation practices including *jhum* and in ensuring freedom to tribal communities to access NTFR. The low-income families are pushed to live under increased economic pressure and as a result the unabated poverty has become inseparable component of ethnic and cultural life of tribal communities. It has been observed that the political, bureaucratic and judicial systems have failed to understand the traditional bond between ethnic and cultural life of indigenous communities and their local eco-system and biodiversity reserve. Instead of making efforts to adapt and incorporate customary laws of those communities, the Central and State governments have tried to impose their policies and laws to justify, define and execute developmental projects and commoditization and commercialization of forest resources. The turnaround of situations is possible. It requires a strong political and bureaucratic determination to undertake certain concrete actions, such as (i) restoration of community landholdings and prohibition of privatization of community and forest lands, (ii) recognizing community forestry and drafting local forest policy, instead of having uniform forest policy for entire country and its regions, (iii) strengthening local community institutions and empowering them in joint forest management mechanisms, (iv) protecting the rights of indigenous people on local forest resources, especially NTFRs, since they heavily depend on forest for livelihood and income, and (v) restoration and promotion of *jhum* cultivation among local communities.

REFERENCES

Bordoloi, B.N. (1986). Problems of Transfer and Alienation of Tribal Land in the Karbi Anglang District of Assam with Special Reference to the Karbis. In B.N. Bordoloi (Ed.) *Alienation of Tribal Land and Indebtedness*. Guwahati: Tribal Research Institute, pp. 121-148.

Bordoloi, B.N. (1999). *Report on Tribal Land Alienation in Assam*. Guwahati: Assam Institute of Research for Tribals and Scheduled Castes.

Borthakur, D.N. (2002). Shifting Cultivation in North-East India: An Approach towards Control. In Bimal J. Deb (Ed.) *Development Priorities in North-East India.* New Delhi: Concept Publishing Company.

D'Souza, Alphonsus. (1999). *North East India.* Jakhama: Kohima Jesuit Region (Mimeo).

D'Souza, Alphonsus. (2001). *Traditional Systems of Forest Conservation in North East India: The Angami Tribe of Nagaland.* Guwahati: North Eastern Social Research Centre (reprint 2005).

Fernandes, W. and Bharali, G. (2002). *The Socio-Economic Situation of Some Tribes in Bishnupur and Palizi.* Guwahati: North Eastern Social Research Centre.

Fernandes, W. and Bharali, G. (2006). *Development-Induced Displacement in Assam 1947-2000: A Qualitative and Quantitative Study of its Extent and Nature*. Guwahati: North Eastern Social Research Centre (mimeo).

Fernandes, W. and Pereira, M. (2005). *Land Relations and Ethnic Conflicts: The Case of North Eastern India*. Guwahati: North Eastern Social Research Centre.

FSI (Forest Survey of India). (2001). *State of Forest Report 2003*. FSI, Ministry of Environment and Forests, Dehradun, India.

Invally, P. (2008). *Water Commodification in Maharashtra: A Case Study of Hindustan Coca Cola Beverage Pvt. Ltd.* Pune: National Centre for Advocacy Studies.

Jamir, A. and A.K.Nongkynrih. (2002). *Understanding Land Ownership and Management Systems of the Khasi-Jayantia and Garo Societies of Meghalaya.* Rome: International Fund for Agriculture Development.

Kumar, A. (2002). A Controversial Eviction Drive. *Frontline*, 19(15).

Nongkynrih, A.K. (2005). The Privatization of Indigenous Community Lands in Meghalaya, *Indigenous Affairs*, 2 (N,5), pp. 45-50.

Mazumdar, D.N. (1987). Land and Land Problems in Garo Hills. In B.B.Dutta and M.N.Karna (Eds.) *Land Relations in North East India*, New Delhi: Peoples Publishing House, pp. 157-166.

Nongkynrih, A.K. (2008). Privatization of Communal Land of the Tribes of North East India: Sociological Viewpoint. In Walter Fernandes and Sanjay Barbora (Ed.) *Land, People and Politics Contest over Tribal Land in Northeast India.* Guwahati: North Eastern Social Research Centre and IWGIA, pp. 16-37.

Pachauri, R. (1990). *Sal Plate Processing and Marketing in West Bengal*, Sustainable Management, Working Paper, 12. New Delhi: Ford Foundation.

Shimray, U.A. (2009). Land Use System in Manipur Hills: A Case Study of the Tangkhul Nagas. In Walter Fernandes and Sanjay Barbora (Ed.) *Land, People and Politics Contest over Tribal Land in Northeast India.* Guwahati: North Eastern Social Research Centre and IWGIA, pp. 88-112.

2

Primary Education in North-East India: The Current Dynamics

Subhashree Sanyal

Primary education has always been considered as one of the core component of human development. Since independence, the Government has made significant investment for promoting the access to primary education even in the remotest part of the country. But when we take into consideration the present scenario of primary education in the North-Eastern region, the progress seems to be a bit sluggish in comparison to the other parts of the country. Researches, reports and other available literature suggest that issues like high drop-outs at the primary level, lack of proper infrastructure in the primary schools, high absenteeism among school teachers, etc., are jeopardizing the efforts of the Government directed towards promoting the level of primary education in the North-Eastern States. The slothful progress of primary education in the region seems to be putting an adverse effect on the overall progress of human development across the North-Eastern States of India.

Background

Being a signatory to the historic declaration of Education for All [EFA] in both Jomtien [1990] and Dakar [2000], India is committed to achieving "total education" across the country. In order to meet these goals, India has initiated the Sarva Shiksha Abhiyan in the year 2000-2001. Simultaneously, the Government in accordance with its constitutional mandate, taken several initiatives in the form of enabling policies, legislations and interventions to spread literacy, promote educational development and bridge gender disparities. In order to achieve this goal, apart from commitment given to the National Policies for Education [NPE], several projects and programmes have been launched in the country during the 1980s and 1990s, namely, Andhra Pradesh Primary Education Project [APPEP], Bihar Education Project [BEP], Lok Jumbish, Shiksha Karmi Projects, District Primary Education Programme [DPEP].

One significant feature of these projects was they never completely covered all the States, in fact the DPEP completely left out all the North-Eastern Projects expect 9 districts of Assam. Therefore, till 2000-01 no efforts have been made in the form of projects and programmes were made to achieve Universal Education in North-Eastern States. It may be noted that due to its distance and also difficult terrains, total coverage for education was never provided until the launching of Sarva Shiksha Abhiyaan [SSA] in 2002-03. SSA for the first time made a serious attempt to universalize free and compulsory education across the country specially the North-Eastern States.

Educational Scenario in North-East India

The fifteenth official census in India was calculated in the year 2011. In a country like India, literacy is the main foundation for social and economic growth. When the British rule ended in India in the year 1947 the literacy rate was just 12 per cent. Over the years, India has changed socially, economically, and globally. After the 2011 Census, literacy rate of India in 2011 was found to be 74.04 per cent. Compared to the adult literacy rate here the youth literacy rate is about 9 per cent higher. Though this seems like a very great accomplishment, it is still a matter of concern that still so many people in India cannot even read and write. The numbers of children who do not get education especially in the rural areas are still high. Though the government has made a law that every child under the age of 14 should get free education, the problem of illiteracy is still at large.

Now, if we consider female literacy rate in India, then it is lower than the male literacy rate as many parents do not allow their female children to go to schools. They get married off at a young age instead. Though child marriage has been lowered to very low levels, it still happens. Many families, especially in rural areas believe that having a male child is better than having a baby girl. So the male child gets all the benefits. Today, the female literacy levels according to the Literacy Rate 2011 Census are 65.46 per cent where the male literacy rate is over 80 per cent. The literacy rate in India has always been a matter of concern but many NGOs initiatives and government advertisement, campaigns and programmes are being held to spread awareness amongst people about the importance of literacy. Also the government has made strict rules for female equality rights. India literacy rate has shown significant rise in the past 10 years. (Census, 2011)

Similarly if we look at the North-Eastern States, there have been disparities in attainment of education across different States.

Literacy Ratio of the North-Eastern States: The Recent Scenario

The following Table 2.1 represents the literacy status of the seven States as well as India in total, male and female ratio as per the 2011 Census.

Table 2.1: Literacy Rates of North-Eastern States [2011] in percentage

Issues/ States	*Assam*	*Meghalaya*	*Mizoram*	*Nagaland*	*Manipur*	*Tripura*	*Arunachal Pradesh*	*Sikkim*	*India*
Overall Literacy	73.18	75.48	91.58	80.11	79.85	87.75	66.95	82.20	74.04
Male Literacy	78. 17	82.40	93.72	83.29	86.49	92.18	73.69	87.29	82.14
Female Literacy	67.27	64.36	89.40	76.69	73.17	83.15	59.57	76.43	65.46

Source: Authors adoption from Census 2011

Assam: Literacy rate in Assam has seen upward trend and is 73.18 per cent as per 2011 population census. Of that, male literacy stands at 78.81 per cent while female literacy is at 67.27 per cent. In

2001, literacy rate in Assam stood at 63.25 per cent of which male and female were 75.23 per cent and 51.85 per cent literate respectively. If we compare the previous census of 2001, the total number of literates was 14,015,354 which has gone up to 19,507,017. Average Literacy rate in Assam for Urban regions was 88.88 per cent in which males were 91.84 per cent literate while female literacy stood at 85.71 per cent. In rural areas of Assam, literacy rate for males and females stood at 76.51 per cent and 64.09 per cent. Average literacy rate in Assam for rural areas was 70.44 per cent. The total number of male literates as per the 2011 Census is 10,756,937 whereas female was 8750,080.

Arunachal Pradesh: Literacy rate in Arunachal Pradesh has seen upward trend and is 66.95 per cent as per 2011 population census. Of that, male literacy stands at 73.69 per cent while female literacy is at 59.57 per cent. In 2001, literacy rate in Arunachal Pradesh stood at 54.34 per cent of which male and female were 65.43 per cent and 40.23 per cent literate respectively. In actual numbers, total literates in Arunachal Pradesh stands at 789,943 of which males were 454,532 and females were 335,411. Average Literacy rate in Arunachal Pradesh for Urban regions was 84.57 per cent in which males were 89.45 per cent literate while female literacy stood at 79.04 per cent. Total literates in urban region of Arunachal Pradesh were 232,838. In rural areas of Arunachal Pradesh, literacy rate for males and females stood at 68.79 per cent and 53.78 per cent. Average literacy rate in Arunachal Pradesh for rural areas was 61.59 per cent. Total literates in rural areas were 557,105.

Meghalaya: Literacy rate in Meghalaya has seen upward trend and is 75.48 per cent as per 2011 population census. Of that, male literacy stands at 77.17 per cent while female literacy is at 73.78 per cent. In 2001, literacy rate in Meghalaya stood at 62.56 per cent of which males and females were 71.18 per cent and 50.43 per cent literate respectively. In actual numbers, total literates in Meghalaya stands at 1,817,761 of which males were 934,091 and females were 883,670. Average Literacy rate in Meghalaya for Urban regions was 91.33 per cent in which males were 93.17 per cent literate while female literacy stood at 89.49 per cent. Total literates in urban regions of Meghalaya were 471,956. In rural areas of Meghalaya, literacy rate for males and females stood at 72.83 per cent and 69.45 per cent. Average literacy rate in Meghalaya for rural areas was 71.15 per cent.

Total literates in rural areas were 1,345,805. Total number of literates has gone up from 1,157,875 in 2001 to 1,817,761 in 2011.

Mizoram: Literacy rate in Mizoram has seen upward trend and is 91.58 per cent as per 2011 population census. Of that, male literacy stands at 93.72 per cent while female literacy is at 89.40 per cent. In 2001, literacy rate in Mizoram stood at 88.80 per cent of which males and females were 92.53 per cent and 86.75 per cent literates respectively. In actual numbers, total literates in Mizoram stands at 847,592 of which males were 438,949 and females were 408,643. Average literacy rate in Mizoram for urban regions was 98.10 per cent in which males were 98.67 per cent literate while female literacy stood at 97.54 per cent. Total literates in urban region of Mizoram were 478,920.In rural areas of Mizoram, literacy rate for males and females stood at 88.35 per cent and 80.04 per cent. Average literacy rate in Mizoram for rural areas was 84.31 per cent. Total literates in rural areas were 368,672. The numbers of literate people have gone up from 661,445 in 2001 to 847,592 in 2011.

Manipur: Literacy rate in Manipur has seen upward trend and is 79.85 per cent as per 2011 population census. Of that, male literacy stands at 86.49 per cent while female literacy is at 73.17 per cent. In 2001, literacy rate in Manipur stood at 70.53 per cent of which males and females were 80.33 per cent and 61.46 per cent literates respectively. In actual numbers, total literates in Manipur stands at 1,891,196 of which males were 1,026,733 and females were 864,463. Average Literacy rate in Manipur for Urban regions was 85.98 per cent in which males were 92.05 per cent literate while female literacy stood at 80.21 per cent. Total literates in urban region of Manipur were 622,315. In rural areas of Manipur, literacy rate for males and females stood at 84.14 per cent and 69.95 per cent. Average literacy rate in Manipur for rural areas was 77.15 per cent. Total literates in rural areas were 1,268,881. The total literacy has gone up from 1,310,534 in 2001 to 1,891,196.

Nagaland: Literacy rate in Nagaland has seen upward trend and is 80.11 per cent as per 2011 population census. Of that, male literacy stands at 83.29 per cent while female literacy is at 76.69 per cent. In 2001, literacy rate in Nagaland stood at 66.59 per cent of which males and females were 76.04 per cent and 56.87 per cent literates respectively. In actual numbers, total literates in Nagaland stands at 1,357,579 of which males were 731,796 and females were 625,783. Average Literacy rate in Nagaland for Urban regions was 90.21 per cent in which males were 92.11 per cent literates while female literacy stood at 88.10 per cent. Total literates in urban regions of Nagaland were 452,780. In rural areas of Nagaland, literacy rate for males and females stood at 79.49 per cent and 72.01 per cent. Average literacy rate in Nagaland for rural areas was 75.86 per cent. Total literates in rural areas were 904,799. The total number of literates has gone up from 1,132,323 in 2001 to 1,357,579 in 2011.

Tripura: Literacy rate in Tripura has seen upward trend and is 87.75 per cent as per 2011 population census. Of that, male literacy stands at 92.18 per cent while female literacy is at 83.15 per cent. In 2001, literacy rate in Tripura stood at 73.19 per cent of which males and females were 82.42 per cent and 64.33 per cent literate respectively. In actual numbers, total literates in Tripura stands at 2,831,742 of which males were 1,515,973 and females were 1,315,769. Average Literacy rate in Tripura for Urban regions was 93.61 per cent in which males were 95.80 per cent literate while female literacy stood at 91.38 per cent. Total literates in urban regions of Tripura were 815,720. In rural areas of Tripura, literacy rate for males and females stood at 90.86 per cent and 80.06 per cent. Average literacy rate in Tripura for rural areas was 85.58 per cent. Total literates in rural areas were 2,016,022.

Sikkim: Literacy rate in Sikkim has seen upward trend and is 82.20 per cent as per 2011 population census. Of that, male literacy stands at 87.29 per cent while female literacy is at 76.43 per cent. In 2001, literacy rate in Sikkim stood at 68.81 per cent of which males and females were 77.38 per cent and 59.63 per cent literate respectively. In actual numbers, total literates in Sikkim stands at 449,294 of which males were 253,364 and females were 195,930. Average Literacy rate in Sikkim for Urban regions was 89.26 per cent in which males were 92.94 per cent literates while female literacy stood at 85.19 per cent. Total literates in urban regions of Sikkim were 122,896. In rural areas of Sikkim, literacy rate for males and females stood at 85.42 per cent and 73.42 per cent. Average literacy rate in Sikkim for rural areas was 79.82 per cent. Total literates in rural areas were 326,398.

Factors affecting Growth of Education in North-East India

Currently a plethora of educational initiatives may be underway in rural India to ensure the access to primary education for every child, but issues of equity and quality still remain the areas of concern.

Education Quality provided by Rural Primary Schools: The schools in rural areas of North Eastern India suffer from infrastructure and human resource deficiencies. Often the topography of rural areas becomes a major deterrent factor and accounts for low attendance of teachers. This discourages parents to send their children specially girls to school. Further lack of teachers and teacher absentism create excess burden on the existing few leading to poor quality of service delivery in rural areas. There is prevalence of several single teacher schools which results in excess burden. Generally teachers refuse to teach in rural areas and those that do are usually under-qualified, therefore, impacting the teaching quality.

High Drop-out Rates especially among Girls: Despite of serious efforts by Government, Civil Societies to integrate the entire population into the Indian education system, large numbers of boys and girls are still without schooling. Frequent absenteeism and irregularity of school teachers since several of them do not stay in the same village as the school Parents' apathy and indifference towards education do not provide enough motivation for their children to attend school. Taboo about girl child education still exists specially in non-tribal States like Tripura and Assam thereby leading to huge drop-out rate among girls. Lack of proper sanitation facilities in schools is another reason discouraging especially adolescent girls from attending schools

Inadequate School Infrastructure: Basic infrastructural components like school building and child friendly classroom, blackboard, teaching learning materials, toilet especially for girls and water facilities, etc., are not available Insufficient and inadequate number of classrooms per school to accommodate all the standards in different rooms and hence system of having multiple standards in the same room at the same time. This hampers the quality teaching and essential learning and reduces the retention of students in school.

Apart from this, getting children to formal educational institutions, ensuring their attendance and continuation until completion of the cycle and adequate quality of education offered, providing non-formal basic education to out-of-school children and young adults, adult literacy and continuing education to the illiterate, drop-out and neo-literates, etc.—still appears to be a daunting challenge, keeping in view the size of target population and limitation on resources—be it financial, human, institutional or organizational.

Children in rural and urban areas continue to be deprived of quality education owing to factors like lack of competent and committed teachers, lack of textbooks or teaching-learning materials, and so on. As a large number of teachers refuse to teach in rural areas, so lack of committed and qualified teachers in rural schools has been instrumental in hampering the quality of primary education; on the other hand, due to lack of proper motivation and encouragement from parents as well as from school teachers, rural children sometimes lose their interest towards pursuing schooling. Besides that, most of the rural schools suffer from non-existent of basic infrastructural facilities like proper classroom buildings, availability of drinking water, electricity-facility, etc. Due to lack of adequate school infrastructure and committed teachers, the rural schools fail to create proper learning atmosphere for rural children which ultimately encourages school drop-outs and poor attendance rate among students. On the other hand, some surveys conducted on rural education reveal that teachers of rural schools receive low income in comparison to their urban counterpart, which often compel them to get involved into additional income-generating activities, ignoring their principal responsibility, i.e., teaching. (Hazra, 2013)

Ensuring a Better Educational Access to North-East India: The Way Forward

Since independence, the Government has taken active initiatives in promoting decentralised planning but it seems to be inadequate in the sphere rural education. In most of the States this idea has not penetrated fully at the grass-roots level. Hazra, 2013 stated that one of the reasons behind the failure to achieve the goal of universalisation of Elementary Education is that the plans which are formulated at higher levels, i.e., national or State levels were quite indifferent about the grass roots dynamics and realities of rural education. On the other hand, cross-country evidences indicate that the formulation of educational policy has not been done properly due to lack of democratic values and influence of powerful elites in administration and very often the policy-implementation process suffers from inefficient policy-administration, corruption, lack of political will and commitment. The rigid ideas of bureaucracy sometimes also hampers in policy implementation process. In States like Assam and Tripura there is high drop-out rate, essential steps are needed towards this direction.

The policy makers should recognize the vicious cycle of illiteracy and inadequacy of education with poverty, one reinforcing the other, both as causes and effects. Therefore, addressing the basic social barriers would help increasing enrolment and retention in schools. The role of civil society is essential for community participation; so their involvement should be sought for promoting awareness amongst parents of drop-out children especially in rural areas of North-East India.

Regularizing and monitoring of teachers attendance is a must requirement along with incentives like quality mid-day meals for ensuring students regular attendance to school. Infrastructure—mainly road and transport facilities are major barriers in this part of the country, the State governments along with the Centre should take essential steps towards development of proper roads in rural and urban areas.

In order to ensure quality education, steps should be taken towards regular recruitments for filling up vacancies, developing a "student friendly curriculum" and emphasis on writing skills rather than on rote learning. To offer an appropriate learning environment, suitable arrangements for separate toilet facilities for girls, availability of drinking water, electricity-facilities, etc., are needed within the school. Drinking water is a major problem in North-East India, therefore, steps can be taken towards ensuring the same to encourage more and more admission in schools. Effective appraisal is mandatory for every scheme implemented towards ensuring Universal Education, hence, appropriate measures should be taken regarding creating a monitoring web, where monitoring ranges from teachers performance, utility of funds, transparency and many other important aspects.

Though the State of education in North-East India has significantly improved but the situation needs improvement especially in the Non-Tribal dominated regions. Steps should be taken towards changing the mindset of people through active community awareness and participation. This will encourage the community to come forward and develop more ownership and acceptance for the programme. Last but not the least, steps can be taken to decentralize the different functioning and monitoring agencies so that more and more participatory democracy and political will can be attained and goal of "Education for All" can be reached and the indicators of Human Development are achieved in their true ethos.

REFERENCES

Government of India (2010). *Selected Educational Statistics 2007-2008.* New Delhi: Department of Education and Statistics Division.

Government of India (2012). *Census of India 2011: Literacy Rates.* Retrieved from *http://censusindia.gov.in/* [Accessed on August 20, 2012]

Hazra, A. (2009). Education Status of Women in India. *Social Welfare.* Vol. 56 (6), pp. 4

Hazra, A. (2013). *India's Social Sector and Millennium Development Goals* Jaipur: Rawat Publications

NEUPA and Department of School Education and Literacy. (2012). *Elementary Education in India: Progress towards UEE.* New Delhi: Government of India

World Bank. (1997). *Primary Education in India.* Washington DC: World Bank

Zaidi, S.M.I.A. *Provision and Utilization of Elementary Education Facilities in North East India: What DISE data reveals?* Retrieved on 20.08.2012.

3

Tribal Development in the North-East India: Realities and Constraints

Ajeet Jaiswal

The 'practice' of tribal development started with the British (though some may claim that emperor Asoka was the first one to initiate the process by creating the post of *Anta-mahamatya*—the minister to look after the downtrodden—but nothing is known about the steps taken in this regard.) But they never concealed their motive in following the 'tasks of mercy'. The aim was either to keep the tribals isolated from the 'national' mainstream, thus keeping them away from the menace of freedom-struggle; or by the same process, to convert them to the religion of Christ so that they identify themselves more with the Crown and its interests in the country. Though they were quite successful in achieving their motives, the tribals, because of their inherent psyche of never being subservient, at will blasted into minor or major revolts. Upto 1920s our national leaders had no inclination about the plight of these 'sons of the soil'[1].

The gravity of situation was well realized at the dawn of independence and Articles 38 (State to secure a social order for the promotion of welfare of the people) and 46 (Promotion of educational and economic interests of Scheduled Castes, Scheduled Tribes and other weaker sections)[2,3] were incorporated into the Constitution not only for the weaker sections of the society to be protected against exploitation but also to help them in economic development. We must, however, add that the task of development was to be implemented by the same bureaucracy within the same normative structure as was existing in the pre-independence times. There had to be a clash between the formal apathetic attitude of bureaucracy and the humanitarian angle envisaged in the Constitution—the resultant mess, thus created, is understandable[27,28].

In the beginning itself, it was realized that tribals must be 'integrated' into the national mainstream—they should neither be assimilated, nor isolated. Without realising the odds heavily weighing against the STs, the task was approached as a routine work. Some of the STs (mostly the vocal and influential among them) cornered the fruits of tribal development in the early Plans—enhancing their position and further accessibility to resources. The very STs have been gaining substantially during every subsequent Plan. These tribal 'elites' have changed their habitats and life styles and live mostly in the urban centres enjoying the best facilities available. They have got assimilated into rest of the society but have retained the tag of ST to garner the benefits that should have gone to their really needy brethren. They, sitting in their luxurious apartments or five star hotels, harp on the poor state of tribals and claim to be the spokesmen of STs in general[3,4,5].

Against this is a vast ocean of humanity being fed on the promises. These poor STs have, since decades, been looking for the governmental measures while toiling, to survive. These also include the STs who were better off but have reached this sorry state because of the wrong deeds of government or exploitation through private individuals/groups. In the name of tribal development roads were constructed and these "roads of development" turned out to be "roads to exploitation". Through these road money-lenders, forest contractors, government officials, etc. entered the tribal world to claim their own shares of tribal resources. Forest policy was reoriented dissociating tribals from land which is not only their source of livelihood but also a very vital aspect of their cultural and psychological make-up. Gradually, they were uprooted—socially, culturally, economically, politically, ritually, psychologically and physically. Today's tribal has a hazy future, dark future, no future at all.

Even though the preference of Anthropologist to the study of tribal development and the consequent prolific growth of literature on the subject, one has reasons to feel dissatisfied with the status of tribal research in India, which is marked, by and large, by a low level of sophistication[6]. No serious attempts have been made even to conceptualize the term 'tribe', and social scientists have willy-nilly accepted a legalistic definition. For them tribes are those which are included in the Scheduled Tribes list in the Constitution of India. The 414 tribes listed following the 1956 Presidential Notification differ greatly in their habitats, modes of production, degree of isolation, degree of acculturation, level of development, social customs, beliefs and so on[7].

Tribal and Development

Starting with the issue of tribal development we must first try to define the two terms 'tribal' and 'development'. Tribals are the people, who in anthropological literature are characterized generally by majority of the following features: isolated ethnic groups, low density of population, primitive economy, primitive technology, primitive political organization, primitive religion and lack of script of their language. Truly speaking, hardly any of the population groups on the Indian mainland can lay claim of being a 'tribal' group. The tribals in India comprise of the groups which are declared tribals as per an order of the President. In other words these are the Scheduled tribes[4,5,6].

The anthropologist's conception of a tribe as a small, culturally distinct and economically self-sufficient community with a language of its own and an autonomous political organization is utterly inappropriate to the so-called tribal groups in India[7]. Some go to the extent of stating that the concept of a tribe is an anachronism in the present-day world, since there is no criterion to divide humanity into two branches, tribes and non-tribes. In the present day world, smaller, isolated, technologically backward communities have either become extinct or have become part of one or the other greater civilizations of the world[8]. It is unfortunate that, following classical colonial anthropology, Indian anthropologists have depicted tribes as small, self-contained, self-sufficient and autonomous communities practicing subsistence economy with the lack of or limited external trade, in which exploitation and social conflict have no place. They have attempted to demonstrate the non-existence of any differentiation among the tribal population and thus strengthened and propagated the same myth of homogeneity held and propagated by political leaders and populists[9]. Failure to take cognizance of the uneven distribution of assets among tribal households, expansion of trade, political centralization and emergence of an exploiting group based on antagonistic production relationships has been responsible for evolving inappropriate strategies for tribal development. Another superficial approach to the problem of tribal development emanates from equating tribal areas with any other economically backward area and recommending identical packages of measures for their uplift. Tribals, as a class, are viewed as poor[10]; they are described as constituting the matrix of Indian poverty[11]. Quite apart from the fact that scantiness in some tribal societies, particularly those living in inaccessible regions, may just represent a mode of living in their natural setting, rather than being reflective of their poverty,[12] the approach oversimplifies the complex problem of tribal development by making it a purely economic one. The problem is more basic and includes, apart from economic development, preservation of ethnic identity, ecology, language, culture, style of living, indigenous practices, etc. A mere plan of economic development would be utterly inadequate. Along with economic planning, there should be social and political planning in an integrated manner.[13] Tribes in India are not only numerous, but also differ widely in their habitat, level of development, modes of production, exposure to the wider world, traditional values, customs, beliefs, etc. There are tribes living in inaccessible hill tops, having minimal contact with the world beyond them. There are tribes in the plains living with non-tribal population and obviously having a large degree of interdependence. There are tribes practicing diverse modes of production, right from hunting, fishing, fruit-gathering to being engaged as industrial-urban workers. Some tribes have gone far ahead of others educationally. There are tribes with collective ownership of land and forest resources ensuring an egalitarian and unstructured social set-up and exhibiting a strong sense of solidarity. There are also tribes having individual ownership of property leading to a structured society akin to our own. While there are tribes which have been coming into the fold of the Hindu cultural pattern, there are those which are moving in the opposite direction.[14] The movement from the tribal to the peasant has not been a unidirectional one[14]. With such diversities, attempts to evolve a general scheme of tribal development, having universal application to all tribes in India, are bound to be abortive. Tribal development, because of the diverse situations, has to be area-specific.

Defining 'development' is quite a tricky preposition. There is no consensus on the definition because different sciences interpret it differently. In layman terms, development is a change in the positive or progress direction. But these terms 'positive' and 'progressive' are value loaded. We should consider development as "Change in the desired direction". The connotation may not much be disputed if one is talking about his/her development. However, when we talk about as sensitive an issue as tribal development, the question that automatically comes is: whose desires—those of the planners, of implementers or of tribals themselves. The question becomes more relevant when we realize that in the name of 'tribal development' others have tried and have imposed their will on tribals[5, 15].

There is another dimension of development, especially when we talk about economic development, because without any proper time frame economic development is meaningless. In full, therefore, development should be defined as the "change in desired direction within stipulated time." And it is a well known fact that we have neither cared for the desire of tribals nor adhered to any self-imposed time limit. The problems and prospects of tribal development of north-eastern region of India may be considered along two axes: (a) as ethnic entities, and (b) as status-class[16].

Tribes in the North-East

Diversity is also a characteristic of the tribal population of the north-eastern region of India. However, tribes of this region may be divided broadly into hill-dwelling and valley-dwelling with distinct economic problems.

The economic problems of the valley-dwelling tribes are not basically different from those of the Indian peasantry in general. Because of their long exposure to the national economy, polity and society, they have retained very little of their indigenous economic and political practices and institutions. Save some social practices and physical features, they are indistinguishable in their modes of production and pattern of consumption from the non-tribal peasant class. Money has made inroads into these societies. Division of labour on the basis of skill has taken place.

Private ownership of resources has taken firm roots resulting in social stratification. Because of non-insulation from stronger national forces, the areas inhabited by plains tribes have become, by and large, more impoverished economically. Barring the case of the elite, members of which are extremely limited in number, the great majority of the plains tribal population has been reduced to a position of poor peasants. As a class, they probably represent a classic case of poverty within poverty. As geographical areas, they are grossly deficient in infrastructural facilities. These tribes suffer from land alienation, landlessness, land fragmentation and outmoded agricultural practices[1,3,5].

Table 3.1: Comparison of Scheduled Tribes Population of North- Eastern States of India: 1991 and 2001

(in Lakhs)

S. No.	India/North Eastern State	1991 Census			2001 Census		
		Total	ST	ST Population Percentage of Total	Total	ST	ST Population Percentage of Total
1.	India	8443.24	806.64	9.55	10286.10	843.26	8.20
2.	Assam	222.94	32.68	14.65	266.56	33.09	12.41
3.	Manipur	18.37	5.80	31.75	21.67	7.41	34.20
4.	Meghalaya	17.61	16.08	91.32	23.19	19.93	85.94
5.	Mizoram	6.86	6.50	94.75	8.89	8.39	94.46
6.	Nagaland	12.16	9.72	79.93	19.90	17.74	89.15
7.	Sikkim	4.06	1.09	26.85	5.41	1.11	20.60
8.	Tripura	27.45	8.73	31.80	31.99	9.93	31.05
9.	Arunachal Pradesh	8.58	6.59	76.80	10.98	7.05	64.22

Source: Ministry of Tribal Affairs, GOI, Annual Report, 2009-10[17].

Here the strategy for tribal development must be both area-specific and household-specific. Apart from creating additional infrastructural facilities like transport and communication, education, health and hygiene, irrigation, etc., specific family-oriented schemes for the uplift of those living below the poverty line must be initiated. Utmost importance should, however, be placed on the formation and development of local skills so that the tribals can diversify their occupations and partake increasingly of the benefits of national developmental measures. Legislative measures granting them some special concessions can also be of some help, although too much of protection for too long a period may prove detrimental to them in the long run[1,4].

Tribes living in the hills, with some minor exceptions, are not yet integrated into the national, economy. Their contacts with the national economy have been minimal largely because of the difficult terrain of their habitation and partly because of deliberate policy measures. Only recently they have been exposed to the national economic system to some extent. Attempts to extend the national economy to these communities in haste in the name of tribal development are likely to be counter-productive of development, besides setting in motion waves of social unrest. In fact, the emergent tribal identity or the recent trend towards retribalization can be attributed to a large extent to their sudden exposure to the national economic forces. Let me elaborate this at some length.

The national economy of India is money controlled and dominated substantially by mercantile and industrial capital. It has links with the world economy, which, in its turn, is dominated by multinationals of industrial capitalist countries. The multinationals have their stronghold in all the developing countries, including India. Being a money economy with private ownership of the means of production, individualism ailed profit motive are the mainsprings of Indian national economy [2,5]. The tribal communities, on the other hand, are far from being fully monetized. A large volume of transactions still takes place within these communities through barter. Money as a medium of exchange is used mostly to settle transactions with the non-tribals, which are analogous to external trade. Money has not emerged in the tribal communities as a store of value (status symbol), measure of value and standard of deferred payments. Although periodic markets exist in these societies where small quantities are bought and sold, often without the use of money in face-to-face transactions at the market site, the market mechanism as the resource allocative device has not developed. A market for factors of production has not come up. In fact, the entire tribal setting is different[5,18,19].

Production is meant primarily for self-consumption, not for transactions. All members of a tribal family, including children, take an active part in production. The technology used is indigenous which fits well into the ecological surrounding. There is a complete absence of machine technology. Collective endeavour is a strongly developed feature in such a society, which can be observed in many economic activities. Here the tribals live in a world of their own. Life is one of great physical hardship, but the tribal life in the hills is also one of gaiety and merriment. Being an insulated society, a tribal community hardly produces any surplus over what is required for self-consumption. The sense of solidarity and strong community feelings preclude anything like poverty in such societies.

It is these tribal societies which are expressing their agonies of transformation with the entry into the national economy. Development, to a non-tribal, means more roads, more industrial units, more production and consumption, more savings and investments. But when roads (railway lines) are constructed linking tile long isolated hill tribal areas, more people come from outside, not for sight-seeing, but for gaining economic advantages. There is land grabbing; tribal land gets alienated. The tribals develop a taste for industrial products[5]. Not having the means to get their wishes fulfilled, they get indebted to the traders and moneylenders, who begin to gain effective control over the tribal resources. Whenever an industrial unit is set up in a tribal area, there is again an inflow of people from other areas to man it, since local expertise for this kind of work is lacking. The tribals uprooted from their soil, become, at the most, wage earners in the factory. The national market is extended. Money takes a firmer grip, making the earlier style of life, indigenous economic, social and political institutions irrelevant. There is even a threat of the tribal population becoming a minority. The fear of losing their ethnic identity looms large on the horizon. If development entails turning the masters into slaves to some modern gadgets, will it not be wiser not to extend the national developmental measures to these tribal societies by legally insulating them from the rest of the national economy? The answer is in the negative. First, the hill tribal areas constitute a segment, and a very important segment, of the national economy. It is here that the mineral, forest and water resources are located and national economic development demands utilization of these resources[18,19]. However, haste must be avoided. The pace and intensity of utilization of resources must be in a manner which produces the least adverse effects on the tribal societies. Secondly, measures to insulate these societies are not going to work for long.

Capitalism represents too strong a force to be contained by legislative devices. Getting themselves integrated into the larger economic order appears, therefore, to be the only way out even in the interest of the tribals themselves. What should be emphasized in the development strategy for hill tribal areas is softening the impact of the changeover and preparing the tribals to accept the challenge with greater ability and vigour. Some of the indigenous institutions will become irrelevant; they will die down. Some will be adjusted to the new economic order and some retained as they are.[20]

The strategy of tribal development also requires defining in clear terms the contents of development for the tribals. These are bound to be different from the national contents. Economic development for the tribals, and also for the North-Eastern region, as I have defined elsewhere, it is a persistent rise in per capita income in real terms emanating from increased domestic factor productivity without accentuating economic disparities. This must be achieved while minimizing the adverse effects on future resource availability and ecology and without jeopardizing, ethnic identity[21].

Development, whether tribal or otherwise, is a value loaded term as it signifies a process of change in the desired direction. It is an indivisible whole, although for the sake of convenience it is decomposed into economic, political, cultural, educational, spiritual and the like. It is a harmonious, balanced progress in different spheres of individual and societal life. Lopsided progress in one sphere is often at the expense of progress in another (others). When this happens, overall development, which we want to maximize, becomes a casualty.

Barriers or Constraints to Tribal Development

North-East India is the meeting point of South Asia, South-East Asia and East Asia. Strategic, political, economic and ideological developments in each of these subcontinents have bearing on one or the other part of North-East India. Besides many of the tribes in North-East India have their counterparts in the adjoining countries. Policy planners in North-East India will require having awareness and sensitivity about the happenings in the neighbouring countries of the three sub-continents of Asia.

Most of the tribal communities of North-East India have traditions of migration from East or South-East Asia, either directly or through stages. In post-independence period during all crisis situations, one or the other interest-group in India and outside has tried to take up these grey areas of tradition, to weaken national cohesion. In the face of it, and in the context of the fact that the tribal communities tend to constitute marginalised status-class, national unity in North-East India can be cemented only by emphasising the significance of India as the vanguard nation against colonialism and as one committed to non-capitalist path of development. There must, however, be consistent and continuous manifestation of these. Ideological dimension must be there in addition to emotional and material dimensions. The density of population per sq km in north-east India was 151 compared to 324 at the all India level during 2001 census. But these figures do not represent the whole of reality. In north-east India as a whole, particularly in Arunachal, practically the entire indigenous population lives on agriculture. On the other hand, Net sown area (NSA) in the region has increased from 31.61 lakh ha in 1977 to 38.05 lakh ha in 1995-96. The proportion of net sown area to total area is as low as 15 per cent as against 46.6 per cent for all-India average. Thus population per sq km of gross cultivated area was much higher in north-east India (587) than that in India as a whole (398)[21]. This reality is to be kept in view in determining population policy and in habitat planning and production planning. In Manipur and Tripura, it is claimed that the King was the owner of all lands in the property sense. Hence, it is asserted that with accession of these princes' states to India, all lands other than those on which personal ownership rights can be established are owned by the state. Even the Law Research Institute of Guahati High Court has accepted these claims without further scrutiny. A careful study of the histories of Manipur and Tripura shows that these claims are not based on incontrovertible facts. On the other hand, land policy of virtual expropriation, based on such claims tends to stiffle participatory development process and rationalise bureaucratic intervention. There are two approaches to law—one is legal positivism which considers 'command law' of the political authority as the source of legitimisation; the other is legal pluralism which accepts the law in terms of which the people live their life as the source of legitimisation. During the colonial dispensation the command law of the rulers prevailed. Now there is a growing awareness even among a section of the judiciary, about the incongruity of the command law orientation in a democratic society. Unfortunately, the initiative taken by the North East Council for appraisal of the legal systems in the different States, particularly in matters relating to land and land-based resources have tended to augment the legal positivistic orientation, notwithstanding some advocacy in favour of the legal pluralism in dealing with inheritance, succession, trots and minor crimes. This will be a source of trouble in future.

Land reforms in India, correctly envisages abolition of exploitative intermediaries between the state and the individuals. But in North-East India most of the tribal communities derive only a part of their income from cultivation. They also earn substantially through collection of roots, tubers, leaves and other jungle products. They have access to these resources as members of their respective communities. Hence, if on the analogy with other areas, community as 'intermediary' between the State and the individual householders is removed, the tribal households will be actually dispossessed of much larger resources, though their rights over small areas of settled agricultural land will be consolidated.[22] The current practice in some parts of the country is to confine cadastral survey to nine degree slope only, because of technical difficulty of accurate survey, beyond that at a reasonable cost. This has adversely affected the rights of tribal landholders in those areas. The exact position in North-East India is to be checked. As against the targeted expenditure of Rs. 154.00 crore in the North-Eastern Region, the actual expenditure was only Rs. 134.19 crore (87.1%). With the aim of creating cultural awareness in the North-Eastern Region and identifying/promoting vanishing folk art traditions in rural/semi-urban areas the North-East Community Support Group (NECSG), North-Eastern Zillmere Community Centre (ZCC) has been set up at Dimapur. The progress under the scheme of Multipurpose Cultural Complexes (MPCC) has been slow and the scheme not yet been evaluated in any of the States. The MPCC did not meet any criteria for a CSS and only about 25 per cent of the projects sanctioned have been completed. As per the Zero Based Budgeting (ZBB) exercise, the scheme was discontinued in Budget 2007–08. Initiatives were taken to set up the Central Institute of Himalayan Cultural Studies at Dahung AP for promoting traditional Buddhist Studies[24,29].

The approach towards utilization of earmarked funds, save for few activities listed above, was far from satisfactory. It is essential that the 10 per cent earmarked resources are not only invested for the North-East States but also in the North-Eastern region. It is obvious that the statistical parameters will have to be suitably modified to reflect the realities of the region and planning strategy will have to be adjusted accordingly. Land-man ratio in most areas of North-East India is such that by reorienting production schedule many of the basic needs can be satisfied locally. If after ensuring that, massive commercialisation is attempted, the population will be less vulnerable to market manipulation by speculators and monopoly concerns. But in many areas an opposite strategy has been adopted and development outlay has been planned accordingly. Thus not infrequently, 'development' is suspected to have provided ability for siphoning of resources from the region[29]. National Committee on Development of Backward Areas has recommended individualisation of communal ownership in the North-East for the sake of 'progress'. As a result, attempts are made to implement development programmes by short circuiting the community. But only powerful individuals within the community can do it, with the support of bureaucracy. This is creating neo-feudal rights where such rights do not exist or exist only in incipient form. In terms of political economy what is taking place is anti-development, through 'development'. The dimensions of constraint indicated in earlier section do not exhaust the list. But these suggest the need of reorientation of approach. In that case it will be found that the poverty alleviation programme as implemented in North-East India is artificial in content; it will also bring out the fact that the schemes meant for the poor are not always compatible with their socio-cultural specificities.[23] Besides this will highlight the problem of balanced operation of various sources of institutional finance.[24,29] Lastly it will give an indication of ancillary legislation for the community organizations to function as channels for the flow of institutional finance.

Acknowledgements

The author grateful for the assistance and encouragement of my supervisor Prof. A.K. Kapoor and co-supervisor Prof. Satwanti Kapoor Department of Anthropology, University of Delhi and Dr. Sahin Sultana, Department of Social Work, Pondicherry University. The author is also thankful to Department of Anthropology and Department of Social Work, Pondicherry University for giving all types of assistance to carry out the present work.

REFERENCES

1. Singh, P. K. (2004). Tribal Development in India: What Went Wrong. *J. Soci. Res.*, 26: I & II, 26-32.
2. Vidyarthi, L. P. and Rai B. K. (1985). *The Tribal Culture in India*. New Delhi: Concept Publishing Company, New Delhi, 24-39
3. Rath, G. C. (2006). *The Tribal Development in India*. New Delhi: Sage Publication. 12-18.
4. Chaudhuri, B. (1982). *Tribal Development in India*. New Delhi: Inter-India Publications.
5. Chaudhuri, B. (1992). *Tribal Development in India: Economy and Agrarian Issues*. New Delhi: Inter-India Publications.

6. Patliy, J. (1974). An Outline of Modes of Production in Tribal India, in Buddhadeb Chaudhury (Ed.): *Tribal Development in India*. New Delhi: Inter-India Publications.
7. Bhagabati, A.C. (2001). *Emergent Tribal Identity in North-East India*. New Delhi: Concept Publishing Company.
8. Majumdar, D.N. (1982). An Appraisal of the Tribal Situation in North East India, In Pankaj Thakur (Ed.), *India's North East*. Tinsukia: Prakash Publishing House.
9. Pathy, Jaganath. (1974). An Outline of Modes of Production in Tribal India, In Buddhadeb Chaudhury (Ed.): *Tribal Development in India*. New Delhi, Inter-India Publications.
10. National Council of Applied Economic Research (1978). *Report on Socio-Economic Conditions of Primitive Tribes in Madhya Pradesh*. New Delhi: National Council of Applied Economic Research
11. Vidyarthi, L.P. (1974). Problems and Prospects of Tribal Development in India, in Buddhadeb Chaudhury (Ed.), *Tribal Development in India*. New Delhi: Inter-India Publications.
12. Sharma, B.D. (1978). *Tribal Development: The Concept and the Frame*. New Delhi: Prachi Prakashan.
13. Burman, B.K.R. (1972). "Integrated Area Approach to the Problems of the Hill Tribes of North East India", in K Singh, (ed), *Tribal Situation in India*. New Delhi: Inter-India Publications.
14. Sinha, S. (1974). Rethinking about Tribes and Indian Civilisation, in Buddhadeb Chaudhury (Ed.), *Tribal Development in India*. New Delhi: Inter-India Publications.
15. Burman, B.K.R. (1989). Problems and Prospects of Tribal Development in North-East India. *Economic and Political Weekly*. Vol. 24 (13), pp. 693-697.
16. Ministry of Tribal Affairs (2010). *Annual Report 2009-10*. New Delhi: Government of India,
17. Pfeffer G.; D.K. Behera (1999). *Contemporary Society*: Tribal Studies (Volume-3), New Delhi: Concept Publishing Company.
18. Pfeffer G.; D. K. Behera (2002). *Contemporary Society*: Tribal Studies. (Volume-9), New Delhi: Concept Publishing Company, New Delhi.
19. Planning Commission (2011). *Annual Report 2010-11*. New Delhi: Government of India.
20. Goswami, A. (1984). Tribal Development with Special Reference to North-East India. *Social Scientist*,Vol.12 (8), pp. 55-60.
21. Burman B. K. R. (1988). Land Reforms in North-East India. *Mainstream*, Vol.1 (8).
22. Burman, B. K. R. (1989). *Towards Poverty Alleviation Programme in Nagaland and Manipur*. New Delhi: Mittal Publishers.

23. Burman, B. K. R. (1990). Problem of Institutional Finance in Northeast India. *Mainstream*. Vol. 3 (5)
24. Ministry of Tribal Affairs (2011). *Annual Report 2010-11*. New Delhi: Government of India.
25. Mehta P. C. (2004). *Ethnographic Atlas of Indian Tribes*. New Delhi: Discovery Publishing House.
26. Sachchidananda, R.R.P. (1996). *Encyclopaedic Profile of Indian Tribes*. New Delhi: Discovery Publishing House.
27. Vaid, N.K. (1992). *Who Care for Tribal Development*. New Delhi: Ashtam Publication
28. Hooja, M. (2004). *Policies and Strategies for Tribal Development: Focus on the Central Tribal Belt*. New Delhi: Rawat Publications.
29. Thakur D. (1994). *Tribal Life in India: Tribal Agriculture and Animal Husbandry*. New Delhi: Deep & Deep Publications.

4

Status of Women in North-East India: An Analysis

Molankal Gangabhushan and *Baiakmen S. Hek*

There is a widespread belief that women in North-East of India, enjoy greater mobility and visibility than women of other communities in the country. This belief based on the fact that the practices such as dowry and bride burning are not very prevalent in the region. This is often cited to portray a picture of equity between men and women in the region and has given rise to the presumption that violence against women is not a major concern in the area. However, women from North-East region are also not free from the implication of violence. The ongoing armed-conflict situation prevalent in the North-East of India has intensified the violence faced by women, which takes the form of sexual, mental or physical abuse, killings and clashes. Although all the members of communities are affected by the armed conflict, the impact on women and girls is far greater because of their status in society and their sex.

Introduction

Women's status depends mainly on their rights and privileges and the roles assigned to them, most often on the basis of gender. Status is determined to a great extent in terms of socio-economic indicators such as income, property, education and skills that open up opportunities of employment, better health and the ability to determine important events in life even when others are opposed to them. These indices are also vitally interlinked with the concepts of power and position. It is widely acknowledged today that one cannot hope for gender equity unless women have a share in the decision-making process in the family and in the public sphere, thus enabling them to access the rights and opportunities provided to them by the state, society and socio-cultural institutions.

Gender inequalities are deeply inherent in the North-Eastern region as their counterparts elsewhere. Gender has always been a basic differentiating factor, as is reflected in the cultural norms that have assigned specific roles to men and women in society, there are very strict gender rules and norms that define the roles, responsibilities and attributes allotted to women. It is these rules again that give rise to the rights and privileges that women have. Many of them may remain 'norms' which women are expected to follow; some, however, get the status of 'customary laws', which have a legal bearing. Like their counterparts of other regions, women of North-Eastern Region has also encountered with a number of specific problems over ages. So in order to understand the status of women in the region it is important to look into those parameters which have created discrimination to women.

North-Eastern India comprising the eight States of Arunachal Pradesh, Assam, Manipur, Meghalaya, Mizoram, Nagaland, Tripura, and Sikkim, is the abode of a large number of tribal and other ethnic groups. The status of women in the North-East is comparatively better than women in the rest of India, both in the tribal as well as the non-tribal societies of the region. They have greater levels of mobility, economic autonomy and control over their labour than many women in other parts of India. Practices such as dowry and bride burning are not prevalent in the region. This is often cited to portray a picture of equity between men and women in the region and has given rise to the presumption that violence against women is not a major concern in the area. Unfortunately that is not the case. Although, tribal societies throughout the world are celebrating for their egalitarian treatment of women, it would be more productive to focus on the changes in their roles. If we examine customary laws regarding women's right and day-to-day practice, it is clear that women are far from equal (Trivedi, 2011).

Gender equality, is a constituent of development as well as an instrument of development. It is an instrument of development because without gender equality other goals of development will also be difficult to achieve, namely the goals of poverty alleviation, economic growth, environmental sustainability, etc. A natural corollary of ensuring gender equality is the elimination of gender discrimination. Inequalities between girls and boys in access to schooling or adequate health care prove a very serious disadvantage to women and girls and limit their capacity to participate in the benefits of development.

Literacy and Women

Education of women is the foremost requirement in improving the status of women in the society. In one of the interesting study on Women in North-East they opine, that "Education being a social reproductive system and meant to take the person's towards the larger society, would then belong to the man's world. Hence, western education came to their society through the church, men were the first to get its benefits. The trend is visible today in the number of those who have had access to higher to education". (Fernandes and Barbora, 2002). This can be one the reason which increases the gap of literacy between male-female, moreover, the other reason is the patriarchy system prevailing in most of the regions. But throughout the ages changes has taken place in terms of women accessing to education. Referring Table 4.1 on literacy rate the literacy among female has drastically improve, so also the gap between male- female ratio has been reduced over the years. Mizoram top the list in terms of giving equal access to education for women.

Table 4.1: Literacy Rate (Provisional Report Census 2011)

States	*Male*	*Female*
Arunachal Pradesh	73.69	59.57
Assam	78.81	67.27
Manipur	86.49	73.17
Meghalaya	77.17	73.78
Mizoram	93.72	89.40
Nagaland	83.29	76.69
Tripura	92.18	83.15
Sikkim	87.29	76.43

Women and Health

In the North-Eastern region, all the States have poor utilisation of health, moreover, absenteeism of medical practitioners and staffs is a regular phenomena especially in rural areas. On the other side, the entire region of North-East being an area of conflict the worst to suffer are the women, as according to World Health Organisation report almost three-fourth of maternal mortality occurs during the period of delivery or the post partum.

Sex ratio is a very important indicator that reflects the status of women in any given society. It reflects whether she enjoys the right to survival, protection and development. Table 4.2 on sex ratio reveals a great improvement in terms of sex ratio such as Manipur, Meghalaya, Mizoram, Tripura, and Assam and these States have crossed the national average which stands at 940 whereas Nagaland, Arunachal Pradesh and Sikkim ranks below the national average. The reasons for the imbalance in the male-female ratio have numerous studies which proclaim the adverse juvenile sex ratio due to the preference to have male offspring.

Table 4.2: Sex Ratio (Census 2001 and 2011)

States	*2001*	*2011*
India	933	940
Arunachal Pradesh	893	920
Assam	935	954
Manipur	974	987
Meghalaya	972	986
Mizoram	935	975
Nagaland	900	931
Tripura	948	961
Sikkim	875	889

Another important aspect of woman's health is the usage of contraceptives, family planning and its knowledge among the masses. Though the sex ratios are towards the increase, the reason which may be accounted for having high fertility rates is due to the desire to have male children, which contributes to the large family size of most North-East India's households. It is important to take note that the recent Census of 2011 accords Meghalaya as the top State with high fertility rate, this indicates the lack of knowledge on the family planning. Another area which needs special attention is on the rise of

HIV/AIDS in this region which is the result of ignorance of modern contraceptives. NACO in 2006 has called the North-East that spreaded HIV/AIDS and a very emergency situation prevailing with over 50,800 people of the 2 lakhs infected Indians primarily from Nagaland, Manipur, Mizoram and Meghalaya.

But overall comparing to the 2001 Census there has been a great improvement all over the region, this indicates that women have been given equal opportunities in terms of the gender equality. It is also worth mentioning that Tawang a district in Arunachal Pradesh rank second among the top district in sex ratio with 1,005 women.

Gender Roles and Labour

North-eastern women generally do a greater amount of work than the men and also undertaking tasks that were traditionally in the men's domain. As young men and women are getting educated and moving away from farm work, the burden of farming is shifting to older women, with young girls helping with domestic tasks. Yet, as elsewhere in India, there is a widespread perception that the gendering of tasks is 'natural' (biologically-determined) and that men 'naturally' do the heavier, riskier tasks while women (presumed to be weaker in body and mind) to do the light and easy work.

The contribution of women's labour may be recognised but it is not perceived as a cost. In a comparative study of the labour input of two groups in Arunachal, Gangwar and Ramkrishnan (1992: 126) note that the traditional systems 'respond both to human needs and ecological imperatives'. The study concludes: 'If labour costs are computed, then the financial investment in slash and burn agriculture is high. However, their labour output is geared to the maximum utilising family labour. As such, it does not have actual financial costs'. The study does not disaggregate the labour input by gender, but ethnographic accounts (Mishra, 1991).

Women and Economy

Women's contribution in the economic sphere is great and unlike in other parts of the country, women of some of the communities not just contribute towards the economy but also enjoy considerable economic autonomy. North-Eastern women have a visible role in the economic life of their communities because they are responsible for producing and processing food for their families. Physical mobility, the freedom to work and to take certain kinds of decisions seems to give women a sense of self-worth and identity. Yet, throughout the region, women have no substantial property rights. The Khasi, Garo and Jaintia in Meghalaya are matrilineal, tracing inheritance and descent through the female line but authority is vested in the mother's brother. The *tiwa* (formerly called *lulungs*) in Assam use to be matrilineal group, but are now in the process of transformation to partriliny. The *dimasa-kachari* in Assam have a unique form of dual female and male descent, daughters tracing lineage through the mother and sons through the father. All the other communities are monogamous, but polygyny is also practiced, especially among the wealthy. A very few groups are polyandrous. Family structures vary from extended families to nuclear, or may be unique as in the case of the Nishi 'long house'. A Nishi's many wives and their children live together in a literally long house. Each wife having her own hearth and access to independent *jhum* fields where she works to grow food for her own children. All the wives re subordinate to their common husband who pays bride price for each and claims full control over her and the produce of her labour. Bride price is embedded in wifehood; in fact, the *taraon-mishmi* word for marriage, '*miyabraiya*', means 'buying a woman' (Baruah, 1976: 111). However, bride price does not protect women from exploitation within the family. Apatani women, known to be most hardworking, have a high bride price that is compensated by their labouring in their husbands' fields; so married women are afraid to be seen to be at leisure. Mizo women are liberated to work but their earnings belong to men; a widow's place in society is precarious because she is herself considered the property of her son (Krishna, 1998a, 1998b).

Tiplut Nongbri (1998: 223) says that the tribal women's 'greater economic independence and freedom of movement' compared to 'their counterparts in non-tribal societies cannot be disputed', but it is 'naïve to equate this with superior social status'. She argues that a 'closer look would show that gender inequality is not alien to tribal societies but is obscured by their poor economic condition which forces men and women to cooperate and share in joint economic activities'.

Since poverty cannot be shared equally by all members of the family and the women folks bear the major brunt of poverty. Moreover, the traditional system of community ownership itself did not have any special provisions for women to inherit land, but they had related rights to land-use. Besides, even where the traditional land management system is used, women do not have an independent say regarding its buying or selling. It is on account of this system where property rights in the North-Eastern region are being enjoyed by men, with the exception of Meghalaya which follows the matrilineal system whereby traditionally the property is being inherited by the female members but, however, in reality it is the male member who manage the property and hence takes the decision on behalf of the women. At the same time we cannot generalize that all women inherit property and moreover the property inherited by the youngest daughter is the ancestral property. Hence, there is also a problem of landlessness among the women in the matrilineal society. Even in economic accounting the service rendered by the women folks are not being tabulated because all these services are being provided and not converted into monetary terms hence remain outside the purview of economic analysis and thereby ignoring the welfare contribution of the women.

Women and Decision Making

Traditionally tribal societies of the North-East were patriarchal that kept a separation between the family and social spheres. With their control over the resources, it was prerogative of men to take major decisions in society. But the persons controlling the production process, women played a decisive role in the family. In other words tribal women's power generated from the control they exercised over the family economy and work management in their subsistence economy. Their role in the family economy influenced their decision making powers in it and conferred on them a higher social status than in caste societies (Fernandes,1993).

In the process of time these traditionally accepted norms were diluted because of intermixing with the mainstream of India which led to the adoption of the dominant patriarchy system which is male biased. In terms of decision making women in the North-Eastern region still lacks equality in either making or even viewing their opinions. It is important to take note here that Meghalaya in spite of being a matrilineal society where women are often viewed to have equal rights with the men folk, but yet when it comes to decision making women are being debarred from making decisions at the governance which is at the local level because traditionally women are not allowed to take part in decision making at the *dorbar* (Village council).

In other regions also decision making is exclusively a male affair. Hence, there is dismal representation of women in the legislative, autonomous district council or parliament from the entire region. Therefore, the women in the region have to go a long way before they will be accepted as equal partners in decision-making bodies.

Violence against Women

Women have been, in internal war, the targets of sustained and frequently brutal violence committed by both parties of armed conflict. Both the sides often use violence to punish or dominate women believed to be sympathetic to the opposite side. Women have been threatened, raped and murdered (Human Rights Watch 1998). Violence against women is a manifestation of the historically unequal power relations between men and women, which have led to domination over and discrimination against women by men and to the prevention of women's full advancement. Domestic violence against women is a universal problem that negatively affects society as a whole. Data collected by the North-East Network, however, suggests that violence against women, particularly domestic violence, is on the rise in the North-East. *The Sangai Express*, local news daily in Manipur reported on January 15, 2004, that since the last few years, crime against women has been on the rise. There were 20 rape cases recorded in 2001, 14 in 2002 and 18 in 2003. Abduction and kidnapping of women and children were recorded as 62 in 2001, 82 in 2002 and 70 in 2003. Domestic Violence by husbands and in-laws accounted for five cases in 2001, 10 in 2002 and six in 2003. Even though there has been an increase in the official recording of crimes against women, it has been estimated that a majority of the cases still go unreported.

According to official records there were a total number of 1,995 cases of rape in Assam from 1998 to November 2000. From 1999-2001, the number of molestations in Assam was 166. In Manipur there were 33 rape cases recorded from 1988-2000 and there were 66 cases of molestation from1999-2001. Most often even if people are aware of these violations the cases are not registered and there is no redressal or compensation of any kind.

The number of rape cases registered with the Manipur Human Rights Commission are in the year 2000-15, 2001-8 and 2002-5 cases.

The nature of rape and the silence that tends to surround it makes it a particularly difficult human rights violation to investigate.

The high rate of domestic violence in Assam is reflected in a study conducted by the Law Research Institute, Guwahati, covering police stations in the 23 districts in Assam which reported 10,423 registered cases of violence against women (including rape, dowry, molestation and kidnapping) over the past 10 years.

Meghalaya too, in spite of its fame of being a matrilineal society whereby women are believed to be highly respected, yet, ironically, domestic violence plagues this society too. According to the official figures released by the Union Ministry of Women and Child Development, overall crime against women in Meghalaya has shown an increasing trend over the past few years. There were 82 cases of rape in 2007, 19 more than the figure recorded in 2005. The overall crime against women in the tribal dominated hilly State has increased from 131 to 176 in this period. But more alarmingly, Meghalaya is one of the very few States from where the number of cases registered or charge sheeted and accused arrested or convicted in these three years are not officially available. (*The Shillong Times*, 2009.)

The status of women has come under close scrutiny in the State also and many are astounded by the fact that women in Meghalaya have a low status despite the matrilineal society. The view that women in Meghalaya's matrilineal societies were socially well placed, requiring no special attention was found lopsided when the parameters of women's status, women specific development indicators, like political participation and decision making, literacy, employment and access to health-care in Meghalaya did not compare well to those of other States. One reason for this could be negligence of women's needs and the half hearted measures taken from women's development in the State at the initial stage.

Thus, we can say that although women in the North-Eastern region enjoys a high social status and autonomy than compared to caste societies, yet the women in the region are not free from the domestic violence which is a clear violation of human rights.

Armed Conflict and Women

The North-Eastern States of India are a volatile region, with a number of active insurgencies. To counter insurgency, the Central Government deployed para-military troops into the North-East. The region, because of its physical and cultural proximity to Myanmar, China, and Bhutan, has for the most part been ignored by the central government, thereby fuelling the insurgents' demand for development and autonomy. The insurgency and the presence of military and para-military affect the work people do. Militarization affects every aspect of life and the worst sufferers are the women who are affected because of their economic roles and their vulnerability to physical, mental and sexual violence.

The ongoing armed-conflict situation prevalent in the North-East of India has intensified the violence faced by women, which takes the form of sexual, mental or physical abuse, killings and clashes. Although all the members of communities are affected by the armed conflict, the impact on women and girls is far greater because of their status in society and their sex. The region, under the shadow of conflict, has witnessed a resurgence of patriarchal values and norms, which have brought with them new restrictions on the movement of women, the dress they wear and more overtly physical violence such as rape, which is systematically used as a tactic against a particular community. All this is compounded by the long social, economic and psychological trauma of armed conflict.

Armed conflicts have created large number of female-headed households where the men have been conscripted, detained, displaced, have disappeared or are dead. Women invariably have to bear greater responsibility for their children and their elderly relatives, and often the wider community when the men in the family have gone. The very fact that many of the menfolk are absent often heightens the insecurity and danger for the women and children left behind, and accelerates the breakdown of the traditional protection and support mechanisms upon which the community, especially women, have previously relied (ICRC, 2006).

The reasons may be varied but to highlight a few, firstly there have been numerous cases of violence perpetrated by State security forces against local and tribal women which is a clear indication of violation of human rights, for instance the enforcement of the Armed Forces Special Powers Act 1958 (AFSPA) which give omnipotent power to the forces to pursue a suspect and retaliate in the name of maintenance of civil administration. Secondly violation of human rights in the unlawful encroachment of land of the local tribal's in the name of security. Another important reason worth mentioning on the issue of security and human rights is to draw a comparison with respect to the issue of AFSPA whereby the issue has been brought to the limelight only when Kashmir proposed a partial removal of the Act from certain districts of the State. Whereas if we looked at the same issue where Irom Sharmila the lone woman crusader fast for eleven years to bring the issue of the human rights violations of the AFSPA remains invisible till date. This shows the gross negligence on the part of the GOI with respect to the issue of security and human rights.

Conclusion

To conclude, the more egalitarian tribal culture of the North-East region, coupled with the absence of some of the rigid practices of other parts of India, like seclusion (*purdah*) and dowry amongst the non tribal populace of the region, certainly gives the women of the region an edge over other women, in terms of visibility and mobility. Women of this region also bear the brunt of discrimination and exploitation at all stages of their lives and that it touches every aspect of a woman's life violating her rights and freedom. Discrimination against women in situations of armed conflict is further accentuated, as in such situations women's subordination is reinforced by all parties. A major hurdle to the participation of women in decision making is the much revered customary laws and practices. Customary laws have a legal bearing among the tribal communities, though there are some which accord equal status to women, but not all customary laws are gender sensitive. It is manifested from the fact that no traditional institutions of governance accept women. Even where women have come in, the roles assigned to them are either peripheral or figurative.

In the political arena, the major impediments to free and fair political participation of women are marginalisation due to economic disparity, patriarchy and criminalised politics at all levels. The implication if which is seen in the inadequate representation of women in decision-making bodies, which has been a historical reality and a consistent feature of the societal structure, dependent very much on the same factors, which enforce the marginalisation of women in other spheres. Though there are efforts to provide space for women, however, the dichotomy between the system committed to theoretical equality of gender and the political economy with its emphasis on growth and modernisation which sharpens class and gender inequalities, remain unresolved.

Hence, we can say that certain changes have taken place in terms of women status in North-East with the present social change, it is only with time that women in the North-Eastern region will actually get to see the light of living in an egalitarian world which is free from all forms of discrimination and accept them as equal partners.

In order to enhance the status of women, concerted efforts are needed in this region, through enhancing women's active participation in government and control over resources, to determine resource allocation to various public services and determine the developmental strategy. There is an urgent need for revision/change in the customary laws for instance the norms/laws which debar women from participating in the polity, especially in the traditional institutions of governance like the village councils.

Violence against women has to be seen as a public health issue because of the repercussions of its increase in the North-East region as well as the general apathy to such incidences. Strict measures need to be established by the government to safeguard visiting There also the need of acknowledging the invisible contribution made by women to society which goes unaccounted. Attitudinal change is to be brought in the society to promote a culture of recognising the egalitarian status of women.

REFERENCES

Baruah, T. K. (1976): 'The Effects of Bride-price on a Mishmi', *Resarun 1975,* Department of Information and Public Relations, Government of Arunachal Pradesh, Shillong, pp. 110-13.

Chaturvedi, Sonalika (2011). Health Status of Women in the North-East. *Yojana,* Vol. 55, 28-31.

Fernandes, W. and Barbora, S. (2002). *Modernisation and Women Status in North-Eastern India: A Comparative Study of Six Tribes.* Guwahati: North-Eastern Social Research Centre.

Gangwar, A.K. and P.S. Ramakrishnan (1992): 'Agriculture and Animal Husbandry among the Sulungs and Nishis of Arunachal Pradesh' in Walter Fernandez (ed.), *National Development and Tribal Deprivation,* Indian Social Institute, New Delhi, pp. 100-28.

Human Rights Watch (1998): *Global Reports on Women's Human Rights.* New Delhi: Oxford University Press.

ICRC (2006): *Women Facing War*. Geneva: International Committee of the Red Cross (ICRC).

Krishna, S. (1998a): 'Arunachal Pradesh' in M.S. Swaminathan (ed), *Gender Dimensions in Biodiversity Management,* New Delhi: Konark Publishers.

Krishna, S. (1998b): 'Mizoram' in M.S. Swaminathan (ed), *Gender Dimensions in Biodiversity Management.* New Delhi: Konark Publishers.

Lalneihzovi. (2009). *Changing Status of Women in North-Eastern State*. New Delhi: Mittal Publication

Mishra, K. (1991): *Women in a Tribal Community.* New Delhi: Vikas Publishing House.

Nongbri, T. (1994): Gender Relations in Matrilineal Socieites. *Lokayan Bulletin,* 10 (5/6), pp. 79-90.

North-East Network (NEN). "Violence against women in North-East India: An enquiry" Retrieved from http://ncw.nic.in/pdfreports/Violence%20against%20women%20in%20North%20East%20India%20-%20An%20Inquiry.pdf [Accessed on : May 24, 2011]

Sahidullah, T. Fatima. "Status of women in Assam". Retrieved from http://www.assamtribune.com/scripts/details.asp?id=jun1709/edit3 [Accessed on February 23, 2011]

The Shillong Times, October 2nd 2009.

Trivedi, T. (2011). *Women in North-East India.* New Delhi: Jnanada Prakashan.

5

Human Deprivation in Multicultural Context : The Case of Assam State

G. Ram

This chapter brings out the patterns of human deprivation in a multi-cultural society, revealing that in such society, basically, cultural groups, rather than regions and sub-regions, are deprived from human development, however, it is generally perceived that a region or sub-region by its adverse nature deprives its population from human development. It urges to focus on the cultural dynamics of people and governance, stressing that the marginalized cultural groups are historically pushed to the regions and territories which are poor in economic resources and prone to natural calamities, on the one hand, and even in the developed regions or territories such groups generally remain deprived from the resources, on the other. Thus, it is not the region or sub-region which deprives them; rather the dynamics of the cultural matrix coupled with that of the politics, dominance-submission, in a historical process, give rise to deprivation of certain groups in relation to others in a multi-cultural space and, therefore, the phenomenon of deprivation or poverty can be dealt with by adopting target group-specific measures because indiscriminate diffusion of mitigating measures over a matrix of unequal cultural groups can only further enhance the gap and deprivation. The discussion is based on the Assam Human Development Report 2003, prepared by the Government of Assam, to understand deprivation in Assam's multi-cultural society. UNDP's Human Development Report 1997 brought out Human Poverty Index, a composite index of people's deprivation in three basic dimensions of well-being; namely, long and healthy life, ability to acquire knowledge and decent standard of living, without which many of choices are simply not available. In the report human poverty is understood as the extent of deprivation from human development. The Assam Human Development Report has calculated human deprivation by using Human Poverty Index at the level of the State, region and district. However, patterns of distribution of deprivation primarily and closely follow a cultural distribution of various groups in the regions and sub-regions, constituting the State's multicultural population, because the regional variations in distribution of human cultures along with differential natural resources have actually determined the patterns of human development as well as human marginalization or deprivation in the State. Hence, a question arises: What patterns of human deprivation are perceived in the multicultural setting of the Assam State? The question is attempted with reference to the linguistic communities, tea garden communities, tribes, religious groups, rural and urban populations and gender distinctions in Assam's two broad regions, namely, the plains and the hills.

Multicultural Setting of the Assam State

Situated between the latitudes 28°18′ and 24° N and the longitudes 89° 46′ and 97° 4′ E in an area of 78,438 square kilometres (29.91%) of India's North-Eastern region, Assam is bounded by Bhutan and the Himalayan ranges in the North, Arunachal Pradesh in the North-East Nagaland and Manipur in the East, Bangladesh, Tripura and Meghalaya in the South and South-West, and West Bengal in the West. The inter-setting of its territory is broadly composed of two zones, viz., (i) the plains of the Brahmaputra and Barak valleys; and (ii) the hill tracts in the north-eastern and central parts. The eastern and the western part of the Brahmaputra valley are called Upper Assam and Lower Assam respectively while the area in-between is referred to as the Central Assam (Nagaon, Morigaon). The eastern Himalayas and sub-Himalayan regions form the boundary of Assam from the north to the south-east. The sub-Himalayan region joins with the Patkai Hills, inhabited by the semi-Mongoloid tribes such as Bhutias, Akas, Daflas, Abors, Mishing and Mishmis. From the east to the west is the Assam range occupied in its middle by several tribes, namely, Nagas, Jaintias, Khasis and Garos. On the southern part of Assam lie the Chin and Lushai (Mizo) hills, Chittagong hill tracts and Tipperah hill. The Shillong plateau, dividing Assam into two valleys, namely, the Brahmaputra valley and the Barak valley, continues eastward through North Cachar hills to join the Naga hills. The Barak valley in the south of Assam is formed of a small boggy and flat plain, interspersed by isolated low hills and long hills-top. On the other hand, the Brahmaputra valley is a wide alluvial plain, about 720 kilometres, with a number of small rivers flowing down across it from north and south to the great Brahmaputra River. Majority of the tributaries of the Brahmaputra depending on the monsoons for their volume dry up during the winter, overflow their banks in the summer and at times cut fresh channels through the alluvial soil of the valley (Raha and Ghosh, 1998). The two valleys are occupied by caste and plain tribal populations; predominantly Bengali and tea garden populations in the Barak valley and the Assamese and Bodo Kachari populations in the Brahmaputra valley.

Geo-strategic location, natural resources, rich bio-diversity, ethnically heterogeneous population and distinct culture have accorded Assam a unique place in India. Moreover, it is part of india's North-eastern region which is characterized by disquiet for several decades, low agricultural productivity, poor infrastructure, tenuous communications and nascent levels of industrial activity, widening gap with the rest of the country and continuous perception of itself as distant and distinct, and unable to participate in the benefits of an expanding and growing economy. In the decade of economic reforms of the 1990s the pattern of incremental investment has been skewed in favour of regions with relative advantages in infrastructure, communication and transportation access, established environment for entrepreneurship and availability of skilled work force. It accounts for about 2.4 per cent of geographical area and 2.59 per cent of population of India. Its density of population is 340 persons per square kilometre as compared with 324 persons for India as a whole. Its 27 districts (23 districts in 2003) vary considerably in size of territory and population. The district of Karbi Anglong, sprawling in an area of almost 10,500 square kilometres, is nearly eight times as large as Hailakandi district whereas Kamrup, the most densely populated district, has a population density almost fourteen times that of the least densely populated district, North Cachar Hills. Most of the people, seven of every eight persons (87.28%) are living in the rural areas. Of its population 12.89 per cent is classified as that of the

Scheduled Tribes which are dominant in the two hill districts, namely, Karbi Anglong and North Cachar Hills (now renamed as Dimasa Hasao). In Kokrajhar and Dhemaji, they constitute over 41 per cent of the total population. Other significant concentrations of the Scheduled Tribes are in Lakhimpur, Goalpara, Bongaigaon, Nalbari and Darrang districts. Between 1901 and 2001 its population increased by a little over eight fold and outpaced the average rate of growth of population in the rest of the country. Its decadal variation of population has been very much larger than that in India as a whole, especially in the first half of the century. However, in the decade 1991-2001 the decadal variation has declined to a rate less than that of India as a whole. Of the population, 51.76 per cent are males and 48.24 per cent are females. In recent decades, the sex ratio for Assam has improved while it is on decline for India as a whole (Government of Assam 2003:1-11; Census 2001).

Blessed with fertile soil and a climate conducive to agriculture, Assam has the potential to expand its agricultural production manifold. The State possesses an estimated 320 million tonnes of coal reserves, oil and natural gas reserves and a vast, largely untapped, potential for power generation. It is well placed to service the needs and markets of other States of the North-East. A nascent industrial infrastructure, capable of being built upon, already exists in Assam. Its gap of economic development is increasing. At Independence, its per capita income was only 4 per cent less than that of the average for India as a whole whereas, in 1998, the average per capita income for India as a whole was over 1.8 times that of Assam. The relative stagnancy in the growth of income is attributable to inability of the component sectors to grow at a rate on par with that of the rest of India. Industrial diversification and growth has been constrained by the inadequacy and quality of complementary infrastructure, the region's geographical isolation, and the lack of well-developed markets. Lack of traditions of indigenous entrepreneurship and tentativeness of private investment from outside has necessitated a major role for the State. The agriculture sector has not grown at a rate comparable to that of the rest of the country. The overall growth rate since the 1980s has been a little over 2 per cent, not sufficient to generate surpluses for investment, or create purchasing power in the rural sector to provide a market for local industries. The inability to build on the promise offered by the horticulture sector is largely attributable to the deficiencies in complementary investments, in storage, in rural transportation and in market facilities. For the same reasons, fisheries, poultry farming and dairy are yet to acquire the momentum required to raise incomes substantially and sustainably. In the post-1997 period, timber processing and the related downstream industries have suffered setbacks, and economic activity and consequently income and employment are at levels much lower than even a decade ago. In absolute terms, the number of poor people in the State has increased from 7.8 million in 1983 to 9.5 million in 1999-2000. Of the State's population, 36.09 per cent continues to live below the poverty line, a figure appreciably above the national average of 26.10 per cent. Amongst many challenges that confront the State and its people the two most important issues are insurgency and low level of development. Inequities, lack of development, perceived discrimination and lack of opportunities for employment have contributed to discord, strife and insurgency which are rarely conducive to development. There is a mismatch between the resources needed to provide basic services, maintain assets, promote growth and development and create infrastructure, and the resources available to the State. Recognizing the intricate connections of development with governance, equity and justice, transparency and effectiveness, human rights and the rule of law the government has taken measures to encourage the decentralization of power and decision-making, empowering people at the grass roots level through elections to Panchayati Raj institutions (Government of Assam, 2003:12-15).

Methodology for Measuring the Human Deprivation

Human deprivation is measured by using Human Poverty Index (HPI), in which the term 'human poverty' is used to be understood as the extent of deprivation from human development. Over the years the connotation of 'poverty' has undergone changes as a result of many writings emphasizing that poverty consists of deprivation of people on several counts in addition to mere shortfall of income from a specified level, i.e., poverty line, going far beyond its initial concept, now-a-days, called income poverty. Since the mid-1990s human development reports of UNDP have been measuring the extent of human poverty by using different indices for developed, developing and less developed countries across the globe. For the developing and less developed countries the Human Poverty Index (HPI) is based on the following three key components:

(i) Deprivation from living long and healthy life, measured by the probability at birth of not surviving up to the age of 40 years;
(ii) deprivation in knowledge captured by adult illiteracy rate; and
(iii) deprivation from living a decent standard of life, which was used to be captured through the following three sub-components:
  (a) percentage of people without access to safe drinking water,
  (b) percentage of people without access to health service, and
  (c) percentage of moderately and severely underweight children.

In the recent years, these three sub-components have been reduced to the following two parts:

(a) Percentage of people without sustainable access to an improved water source; and
(b) percentage of children underweight for age (UNDP 1990; Government of Assam 2003: 16, 19; Bezbaruah and Saha 2009:54).

Thus, the concept of human poverty goes beyond conventional income poverty and the HPI is intended to focus attention on concerns of developing countries and on groups of people who are 'deprived'. While the Human Development Index (HDI) measures overall progress in a country in achieving human development, the Human Poverty Index (HPI) reflects the distribution of progress and measures the backlog of deprivations. The HPI measures deprivation in the same dimensions of basic human development as the HDI. Yet, like HDI the concept of HPI is not broad enough to include dimensions such as voicelessness and others (Bezbaruah and Saha, 2009:54).

In the Assam Human Development Report 2003 the human poverty index, though broadly calculated on the UNDP's line, is based on the broad UNDP formula, moderated for and adapted to its relevance in the context of the Assam State, due to non-availability of data on some of the components and, at the same time, availability of data on some key indicators such as deprivation from a decent standard of living. Therefore, while deprivation from long and healthy life has been measured under the UNDP approach by considering the probability at birth of people not surviving up to the age 40, deprivation from knowledge was captured by general illiteracy rate and to capture deprivation from decent standard of living there were used the following five indicators:

(i) Percentage of people without access to an improved drinking water source,
(ii) percentage of children underweight at birth,
(iii) percentage of population without access to health services,
(iv) percentage of households in *kachcha* dwellings, and
(v) percentage of households without access to proper place of defecation (Bezbaruah and Saha, 2009:55; Government of Assam, 2003:49).

The first-three indicators for calculation of HPI for Assam are more or less those which have been used by UNDP, however, the data on the second component, namely, percentage of children underweight at birth were not considered very reliable, while the components (iv) and (v) are considered obviously important and objective indicators of deprivation from decent living standard. Therefore, there were prepared two versions of human poverty index, viz., HPI (A) and HPI (B) in the Assam Human Development Report 2003. In HPI (A), deprivation from decent living standard was calculated using the first-three indicators to keep broad parity with UNDP's HPI calculations while, to make the index more relevant to Assam, the HPI (B) version was based on the score calculated from all the five components of deprivation from decent living standard. The estimation of human poverty in the report is based on the data compiled from a sample of 17,140 households in 219 blocks and 52 urban centres in the year 1999. The extensive coverage of the sample survey has been used to construct the Human Poverty Index (HPI) not only for the State but also for its different districts as well as different population segments such as males and females and Scheduled Castes, Scheduled Tribes, Other Backward Classes and General Category (Bezbaruah and Saha 2009:55; Government of Assam, 2003:49).

Incidence of the Human Deprivation in Assam

The report has shown the human deprivation in the State as a whole as well as in its districts. This type of demonstration is useful to fix the deprivation or poverty at district level and to understand it in terms of inter-district variation by using comparative perspective, on the one hand, and at the level of different communities and inter-community variations, on the other hand, by aggregation of the districts into regions on the basis of cultural groups. The following Table 5.1 shows HPI and reveals the incidence of human deprivation in different districts of Assam:

Table 5.1: District-wise Human Poverty Index in Assam, 1999

District	*HPI (A)*		*HPI (B)*	
	Value (in %)	*Rank*	*Value (in %)*	*Rank*
Sibsagar	10.31	1	24.28	1
Dibrugarh	13.98	2	24.30	2
Golaghat	14.52	3	30.84	4
Nalbari	15.63	4	33.04	7
Kamrup	17.44	5	24.72	3
Nagaon	19.16	6	34.06	9
Dhemaji	19.60	7	32.81	15
Lakhimpur	20.23	8	32.81	6
Morigaon	20.28	9	34.06	9
Jorhat	21.94	10	31.33	5
Barpeta	22.83	11	35.22	13
Darrang	23.30	12	34.43	11
Bongaigaon	24.03	13	35.37	14
Sonitpur	24.68	14	34.60	12
Goalpara	26.30	15	39.55	16
Hailakandi	27.00	16	41.02	19
Tinsukia	29.14	17	33.71	8
Cachar	29.22	18	39.71	17
N C Hills	31.44	19	42.66	21
Kokrajhar	31.51	20	42.73	22
Dhubri	31.98	21	41.92	20
Karimganj	33.38	22	40.22	18
Karbi Anglong	33.52	23	44.44	23
ASSAM	23.24	—	34.30	—

Source: Government of Assam (2003).

The human poverty index works out the State average value of 23.24 per cent, by using the HPI (A) version, while by applying the HPI (B) version it estimates a higher value of 34.30 per cent. Though the ranks of different districts by the two HPI versions are not identical there is broad correspondence between the two sets of ranking and, both, the versions are revealing similar patterns of the broad geographical concentration of human deprivation through the district level HPI figures. Under both the HPI versions, eleven districts have an HPI value lower than the average HPI value for Assam as a whole and twelve districts have a higher value than it. Karbi Anglong district has the highest HPI value of 33.52, indicating that this district has the highest number of people in human poverty, while Sivasagar district has the lowest HPI value, demonstrating that this district has the least number of people in human poverty. On the basis of the HPI value the districts broadly form the following two groups:

(i) The western border districts (except Kamrup, Nalbari and Barpeta), the southern districts, the northern districts and the districts of tribal predominance having high HPI values; and
(ii) the north-eastern districts (except Tinsukia) having low HPI values.

Cultural Patterns of the Human Deprivation in Assam

When the data on the incidence of poverty in different districts and regions are looked at from the point of view of cultural variation and distribution of human groups they reveal cultural patterns of human deprivation in the State. Thus, the human phenomenon is viewed amidst the humans, their collectivities, their collective properties and their activities rather than in their physical space. The cultural patterns of human deprivation in the State emerging from the report are as follows:

(i) *The tribal, tea garden and religious and linguistic minority communities have higher incidence of human deprivation*: Human deprivation is more widespread in the frequently flood ravaged and predominantly tribal population inhabited districts, in the hill districts dominated by tribal peoples, in the places occupied by tea garden communities, in the districts having large segments of immigrant Muslims and in the Barak valley predominantly inhabited by the Bengali and tea garden communities.

In general the incidence of deprivation is found in the western and southern parts and in the hill districts, all, forming periphery of the State and mostly occupied by the marginalized people. The western border districts of Dhubri, Kokrajhar and Goalpara are having Muslim and tribal majority population whereas on the other side Bongaigaon, Darrang and Sonitpur are flood-prone areas inhabiting tribal population along the border of Arunachal Pradesh. The southern districts of Karimganj, Hailakandi and Cachar in the Barak valley are dominated by the Bengali (including Muslims) and tea garden populations. The districts of NC Hills and Karbi Anglong are the two districts predominated by tribal populations. The Tinsukia district in the eastern part records high incidence of poverty, perhaps because of high concentration of poor tea garden population amidst the richness of the region.

(ii) *Lower incidence of human deprivation prevails in the other communities:*

In the central and the lower central parts of the Brahmaputra valley HDI values are higher and HPI values are around the State's average. The incidence of human poverty is found much less in the resource-rich districts towards the east.

As one moves eastward from the western border districts of Dhubri, Kokrajhar and Goalpara, the concentration of poverty falls and the district HPI values lie more or less around the overall State HPI; however, the district of Kamrup in the State's western part has even a relatively low HPI value. Kamrup has a high concentration of urban population as the city of Guwahati is the gateway to the entire North-Eastern region and, therefore, the district shows a relatively low incidence of poverty.

Besides, located in the Brahmaputra's fertile bed, the district of Nalbari records an incidence of low poverty. The lowest incidence of poverty is observed in the two districts of Sivasagar and Dibrugarh forming the tea plantation and mineral rich regions of Upper Assam.

(iii) *Women, the Scheduled Tribes and the immigrant Muslims of the western districts are lagging behind with other groups in literacy (knowledge)*

Women, particularly in the Scheduled Castes, and the tribal

groups especially in the hills lag much behind others in literacy. In general, the Scheduled Castes in Assam are not disadvantaged whereas the Other Backward Classes are relatively more advanced in literacy. According to Census 2001 the literacy rate of the Scheduled Castes in Assam is 66.78 per cent. High illiteracy even in the general category of the immigrant Muslims in the poorer western districts of Dhubri, Goalpara and Barpeta seems to reflect widespread illiteracy in the Muslim population of immigrant origin in the State. According to Census 2001 the literacy rate in these three districts is 48.21 per cent, 58.03 per cent and 56.24 per cent respectively against 63.25 per cent for the State as a whole (Government of Assam, 2010:20-21).

(iv) *Women are somewhat less deprived than men in case of longevity*

According to the Census 2001 the longevity of life calculated by life expectancy at birth (LEB) for the males and the females in Assam is 57.1 and 57.6 years respectively. During the period 1990-92, the LEB in Assam was lower for women than for men but from 1992 onwards the LEB for women began to exceed the LEB for men and the trend is still continuing (Government of Assam, 2003:114-15). The longevity status reversal perhaps owes to expanding health facilities as well as health awareness throgh exposure to modern media. It seems that natural and social factors, combined, operate to determine gender differentials of the mortality rates and incidence of morbidity in the age groups. Females have better natural resilience than males but, as compared to males, they suffer neglect and deprivation in respect of resource allocation in society. The preference for male children leads to a higher allocation of nutritional and medical resources for them and to a relative indifference towards female children. Deprivation and neglect of the girl child can lead to greater female infant mortality rate (FIMR). Therefore, higher male neo-natal mortality accounts for the higher IMR for males and natural female resilience survives neglect and deprivation. The effects of early malnutrition and health neglect of girl children account for higher mortality rates (MR) amongst adolescent and growing girls. Poor nutrition also leads to a high incidence of anaemia among pregnant women. Women are trained since childhood in self-abnegation and self-sacrifice and are prepared to endure a lot of needless suffering (Government of Assam, 2003:116-20).

(v) *There is deprivation of one-third population from access to improved drinking water*
The districts predominated by the tribal population and those in the Barak valley predominated by the Bengali and tea garden populations are relatively more deprived from access to improved source of water, probably due to poor ground water endowment (Bezbaruah and Saha, 2009:56).

(vi) *There is more acute deprivation from health care facilities in the district of Karbi Anglong than anywhere else in the State.*
It suggests that the tribal areas are lagging far behind other areas in respect of health facilities.

(vii) *The highest percentage of underweight children is in Dhubri and Tinsukia districts*
Understandably, Dhubri is a poor border district of the immigrant dominated population. But in an economically advanced district like Tinsukia it may be due to some extensive inequality of income and opportunities where certain segments of population, most probably tea garden labourers, are severely deprived from health care services.

(viii) *Scheduled Tribes are more deprived from shelter and sanitation*
Deprivation from shelter and sanitation is very extensive among all segments of the population. Yet, it is more acute among the Scheduled Caste groups than the other groups (Bezbaruah and Saha, 2009:56). This fact points towards low living standard of tribal people in the State.
Thus, the patterns reveal that the resources are more concentrated as well as utilized to the advantage of the forward people in the State, depriving the other groups. The tribal people in the hills are lacking on all counts and acutely deprived from human development in all of its indicators.

Mitigating the Human Deprivation in Assam

Considering the salience of cultural diversities, target specific packages of development for different population groups will be more effective to mitigate human deprivation rather than a uniform intervention package. Deprivation of the tribal, minorities, immigrants, tea garden labour, women and the rural people differs in nature and it cannot be tackled by adopting uniform policies and development progammes for all the groups in the State or its various regions. Rather, the groups should be focused as targets of human development and their participation should be ensured in managing their own affairs of development under the State's supervision. Target group specific and prioritized needs based development programmes will show concrete effects in the life of the groups as a result of concerted use of the resources. The past experience shows that prioritised and group target specific programmes have made difference in the life of the deprived segments like the Scheduled Castes and the Scheduled Tribes in many parts of India after the Independence. If the programmes common for all segments are implemented, then, the scarce resources will be usurped by only the developed segments, widening further the gap between the segments. Even when a uniform programme is implemented among the differently deprived segments it fails to improve their life. From this, it can be derived that within the deprived segments also only 'deprived' should be addressed by the priority based programmes and the 'already benefited' should be excluded so as to keep away the 'entrenched interests' of the segments and diffuse the benefits uniformly.

Conclusion

In a multicultural situation, human deprivation, in real sense, is more cultural than regional. Its source is found both in the cultural conditions of the communities and the distribution of resources in the State's hand. Their culture lacks sufficient means to access to the resources, on the one hand, and the State's resources remain concentrated in some of the population segments, on the other. For empowerment of deprived sections in a multicultural situation special, target specific and exigency packages are required to be implemented.

REFERENCES

Bezbaruah, M. P. and Sanjoy S. (2009). 'Dimensions of human poverty in Assam', in Government of Assam (ed), *Compendium of Research Papers on Human Development: Human Development Experience in Assam, Vol. II.* Dispur: Department of Planning and Development, Government of Assam, pp. 54-56.

Government of Assam. (2003). *Assam Human Development Report 2003.* Dispur, Guwahati: Department of Planning & Development.

Government of Assam. (2010). *Statistical Handbook: Assam 2009.* Guwahati: Directorate of Economics and Statistics.

Government of India. (2001). *Census of India 2001: Assam State.* New Delhi: Directorate of Census.

Raha, M.K. and Ghosh, A.K. (1998). *Northeast India.* New Delhi: Gyan Publishing House.

UNDP. (1990). *Human Development Report 1990.* New York: Oxford University Press.

6

State of Primary Education in Meghalaya: With Special Reference to Mid-Day Meal Scheme

Sumarbin Umdor

Introduction

The provision of universal elementary education for all children in India is a constitutional mandate which is enshrined in Article 45 of the Directive Principles of State Policy according to which 'The State shall endeavour to provide, within a period of ten years from the commencement of this Constitution, for free and compulsory education for all children until they complete the age of fourteen years'. Towards achieving this goal, the National Policy on Education of 1986 had declared that the country shall endeavour to secure free and compulsory quality education to all children up to fourteen years by the turn of the twenty first century. However, till 2001 some of the programmes or projects initiated to achieve the objective of universalization of elementary education (UEE) were state specific and none of these covered any of the North-Eastern State. This situation changed in 2000 when Sarva Siksha Abhiyan (SSA) was launched as a nationwide time bound project to achieve the goal of UEE in the whole of the country. Although SSA was launched in November 2000, only three States in the North-East (Assam, Mizoram, and Nagaland) could start it in 2001–02, Meghalaya, Sikkim and Tripura took it up in 2002-03/2003-04. Arunachal Pradesh and Manipur began in 2004-05.

The Tenth Five Year Plan has laid emphasis on achieving the goal of UEE guided by five parameters, namely (i) Universal Access (ii) Universal Enrolment, (iii) Universal Retention, (iv) Universal Achievement, and (v) Equity. The major schemes of elementary education sector during the Tenth Plan period included SSA, District Primary Education Programme (DPEP), National Programme of Nutritional Support to Primary Education, Teacher Education Scheme, and Kasturba Gandhi Balika Vidyalaya Scheme (KGBVS). In the Eleventh Plan, DPEP and KGBVS have been subsumed under SSA (GoI, 2008). In Meghalaya, the two important programmes under the Eleventh Plan for elementary education are the SSA and National Programme of Nutritional Support to Primary Education also known as mid-day meal schemes (MDMS). These two programmes are centrally sponsored schemes jointly funded by the Centre and State government in the ratio of 90 : 10 respectively. The total outlay for these two schemes during the Eleventh Plan period is 22,510 and 7,100 lakhs respectively (GoM, 2009).

Of the two programmes mentioned in the preceding paragraph, the MDMS or school feeding programme has long been advocated as an effective way of improving both nutritional and education outcomes in developing countries. It is seen as a way to boost school enrolment, attendance cognitive performance rates. It is also one of the measures to reduce high drop-out rates among children in developing countries. School feeding is also meant to address malnourishment and undernourishment among children by providing them meals that have high nutritional values in the schools. Another important objective of such programme is in terms of promotion of social equity in society. In India, the supposedly simple act of sitting together and sharing of meals by children from different caste backgrounds serves to undermine the long entrenched caste prejudices particularly in the rural areas.

There are many studies from several countries which suggest that the improvement in nutrition levels of school age children are positively associated with gains in schooling in many areas such as enrolment at younger ages, less grade repetition, less absenteeism, more grades completed, and better performance on test scores. Studies from India also suggest that the programme facilitate in elimination of classroom hunger and in improvement of enrolment rates, and also in the daily attendance of children attending school. Studies on the MDMS in some States like Delhi, Madhya Pradesh, Tamil Nadu, Karnataka, Rajasthan and Chhattisgarh provide ample evidence that the scheme has had a significant impact in reducing classroom hunger and in increasing enrolment rate and daily attendance of children in school. It has also had a positive impact on school enrolment of female children and those from disadvantage families like the *dalits* and the scheduled tribes (Dreze and Kingdon, 2001; Dreze and Goyal, 2003; Jain and Shah, 2005; Samson, Noronha and De, 2005; Khera, 2006). On the flip side, the MDMS has also put a lot of strain on the teaching process particularly in schools with less number of teachers and few classrooms. Besides, the scheme has also been plagued by other problems such as poor quality of the f ood served, absence of supporting infrastructure such as kitchen, drinking water, storage facilities and utensils and also the lack of proper monitoring and supervision mechanism (Afridi, 2005 and Sinha, 2008).

For Meghalaya, the proper and successful implementation MDMS is of particular importance as this scheme seeks to address the twin issues of chronic malnutrition and high school drop-out rate among

Table 6.1: Nutritional Status of Children and School Drop-out Rate in Meghalaya

Indicators of nutritional status of children				
% of children under 3 years old	*NFHS-2 (1998-99)*		*NFHS-3 (2005-06)*	
	Meghalaya	*India*	*Meghalaya*	*India*
Underweight (too thin for age)	28.6	42.7	42.9	40.4
Stunted (too short for age)	48.8	51.6	47.7	44.9
Wasted (too thin for height)	14.9	19.7	31.8	22.9

School Dropout Rates in Class I-V (%)

Year	*Boys*		*Girls*		*Total*	
	Meghalaya	*India*	*Meghalaya*	*India*	*Meghalaya*	*India*
1998-99	61.07	38.23	63.77	41.34	62.44	39.58
2003-04	53.92	33.74	52.91	28.57	53.42	31.47
2004-05	51.77	30.81	48.15	25.42	49.97	29.00
2005-06	42.81	29.52	41.95	22.50	42.38	26.45

Source: 1. Ministry of Human Resource Development, GoI, Annual Reports (various years)
2. http://www.nfhsindia.org/pdf/Meghalaya.pdf

children, both being major area of concern for the State, the latter being a major impediment in the realizing the goal of UEE. Table 6.1 reveals the unenviable situation occupied by the State in this regard, with the proportion of undernourished children and primary school drop-out rate significantly higher than the all India situation[1].

Given the importance of MDMS in attaining the goals of UEE, this article reports and discusses the results of a survey on the State of implementation of the schemes in primary schools in Meghalaya This article is based on study of the implementation of MDMS in the State, for which field survey was carried out between May and June of 2008. As part of the survey, interviews were conducted with school authority (members of school managing committee/principal/head teacher/teacher in-charge) and students covering 304 primary schools (class I to IV) drawn at random from a database of schools implementing the scheme. Along with school survey, information was also collected from office of DIs[2].

The article is organized into five sections including the introductory section which explains the data sources and methodology. Section II provides a brief background of Meghalaya and the status of primary education in the State. Section III discusses the organizational structure of the MDMS in the State and also provides a brief profile of the sample schools. The ensuing section reports the findings of the survey. Concluding remarks are given at the end.

Brief Profile of Meghalaya and Status of Primary Education in Meghalaya

Meghalaya is a hilly State in the North-East India which is predominantly inhabited by scheduled tribes which account for nearly 86 per cent of the State population. The State is divided seven administrative districts. The eastern part of the State comprising of the four districts of East Khasi Hills, West Khasi Hills, Jaintia Hills and Ri Bhoi is manily inhabited by the *Khasi, Jaintia, Bhoi* and *War*, collectively known as the *Hynniewtrep* people. The western part of Meghalaya constituting the three districts East Garo Hills, West Garo Hills and South Garo Hills is predominantly inhabited by the *Garos*. The tribal State has its own socio-economic characteristics which are presented in Table 6.2.

Table 6.2: Socio-economic Characteristics of Meghalaya

General	
Area (in sq.km.)	22429
Number of Districts	7
Number of Blocks	39
Number of Villages	6026
Demographic (2001):	
Total Population	2318822
Population density per Sq. Km	103
Percentage of rural Population	80.4
Sex ratio	972
Percentage of literacy	62.6
Female literacy rate	59.6
ST population	85.9
Total workers	970146
Total Main and Marginal workers	757011/213135
Percentage of workers in agriculture	65.8
Dependency ratio	884.1
Health indicators:	
Birth Rate (SRS 2007)	24.7
Death Rate (SRS 2007)	8
Infant Mortality Rate (SRS 2006)	53
Infrastructure:	
Road density per 100 Sq. Km (2005-06)	35.57
Percentage of unconnected habitation (2008)	47
Percentage of villages electrified (2001)	44.93
Average population served per bank office/branches (2007)	12467
State Domestic Product:	
NSDP at constant (1999-2000) prices for 2006-07	422483
Per Capita Income at constant (1999-2000) prices for 2004-05	17408

Source: GoI (2001), RHS Bulletin, (2007), Umdor and Panda (2007), Government of Meghalaya (GoM, 2007a & 2007b).

Most of the schools in Meghalaya are run by private bodies. In 2008-09, 67 per cent of the schools belong to this category, while at all India level this category accounted for only 19 per cent of the total schools. Most of the primary schools in the State lack the basic infrastructure and many of them are one teacher (19%) and one room (30%) schools with very minimal facilities. There are 4.6 lakhs children enrolled in primary schools in the State with the percentage of boys and girls enrolment almost equal. While the gross enrolment is very high, the net enrolment ratio, which gives the number of children of appropriate age group actually registered at various levels of schooling, shows that the ratio has improved for the State over the years but is still below the national average of 98.5 (Mehta, 2011). The primary school related indicators are presented in Table 6.3.

Table 6.3: Primary School related indicators

Number of schools	*11467*
Number of children enrolled	4459714
percentage of girls enrolment	50.35
Percentage of primary schools	73.86
Percentage of primary schools by private management	67.09
Percentage of primary schools having single classrooms	30.39
Percentage of single teacher primary school	18.05
Percentage of primary schools having toilet	28.21
Percentage of primary schools having kitchen shed (Govt. and govt. aided)	25.86
Gender Parity Index	1.01
Pupil teacher ratio	16
Gross Enrolment ratio: Primary	193.08
Net Enrolment ratio for primary school (for 2004-05)	93.01

Source: Elementary Education in India, Progress towards UEE (Mehta, 2011)

Organizational Structure and Logistics of the MDMS and Profile on Sample Schools

The organizational structure and logistics for implementing the MDMS in the State is as follows. At the State level, there is a steering cum-monitoring committee which is the apex decision making body concerning the implementing of the scheme, while the Directorate of Elementary and Mass Education under the Department of Education of the State government is the main agency for overseeing the implementing of the scheme.

A similar structure exists at the districts, with a steering-cum-monitoring committee headed by the Deputy Commissioner of the district as the Chairman and other government officials of the district as members. The main department responsible for implementing the scheme in the districts is the office of the Deputy Inspector of school (DI). The office of DI is responsible for the release and disbursement of foodgrains and cooking costs to schools under its jurisdictions. It is also in charge of supervising and monitoring the functioning of the scheme ensuring that the schools implements the scheme is as per norms. At the school level, the school management committee and the village education committee under different cluster resource centre are responsibility for implementation of the scheme.

The process of delivery of foodgrains to schools starts as follows. Allotment and lifting report of foodgrains against each dealer of respective block is made by the Department of Food & Civil Supplies of the State government. Foodgrains are issued by Food Corporation of India to dealer based on allocation made by the above department. After the lifting report of foodgrains is received by the office of DIs, a distribution list to all eligible schools is made as per latest enrolment. The wholesalers/dealers issue the foodgrains to the schools based on the enrolment of children as confirmed by the office of DIs.

The process of release of cooking costs to the schools starts with the Planning and Finance departments of the State government giving the necessary approvals for appropriating funds released by the Government of India under the scheme. The funds are then released to respective Deputy Commissioners of districts who in turn release the amount due to the DIs at each sub-divisional level for disbursement to schools under their jurisdiction.

The article is based on study of the implementation of MDMS in the State. Of the sample schools, a little over half are government schools and 41 per cent government aided schools. As much as 96 per cent of the schools surveyed are schools having classes from pre-primary up to primary level (up to class IV), with the remaining schools having classes above primary level. Also, 80 per cent of the schools are running only morning shift having two to three hours classes generally between 6 - 7 a.m. and 9 - 10 a.m. during school days. The other schools are day shift schools. Two schools are found to be running both morning and day sections.

In terms of infrastructure, a little over 25 per cent of the sample schools are one-room schools and near about the same proportion of sample schools are having two rooms only. Thus more than half of the schools covered in the survey are either one or two-room schools. In terms of number of teachers, 18 per cent of the sample schools have only one teacher and about 46 per cent are having two teachers. The percentage of sample schools having three or more than three teachers are 18 per cent and 21 per cent respectively (Table 6.4).

State of Implementation of MDMS in Meghalaya

Following the Supreme Court Direction of 2001, Meghalaya has been

Table 6.4: Basic Information on Sample Schools

Districts/State	East Khasi Hills	West Khasi Hills	Jaintia Hills	Ri Bhoi	West Garo Hills	East Garo Hills	South Garo Hills	State	P—as percentage of State
Number of schools	58	65	34	20	62	44	21	304	100
Type of schools									
(a) Government	17	20	18	8	49	27	20	**159**	**52.3**
(b) Government aided	40	39	14	10	12	10	1	**126**	**41.4**
(c) Others (EGS, Private)	1	6	2	2	1	7	0	**19**	**6.3**
Number of classes									
(a) Preprimary to primary	53	62	32	18	62	44	21	**292**	**96.1**
(b) Preprimary to above preprimary	5	3	2	2	—	—	—	**12**	**3.9**
Schools hours									
(a) Morning shift	42	55	24	18	53	31	19	**242**	**79.6**
(b) Day shift	15	10	10	1	9	13	2	**60**	**19.7**
(c) Both	1	—	—	1	—	—	—	**2**	**0.7**
Number of classrooms									
(a) One	9	14	12	7	25	11	-	**78**	**25.7**
(b) Two	12	11	6	3	22	16	10	**80**	**26.3**
(c) Three	16	12	7	3	10	10	8	**66**	**21.7**
(c) More than three rooms	21	28	9	7	5	7	3	**80**	**26.3**
Number of teachers									
(d) One	5	9	3	1	27	7	4	**56**	**18.4**
(e) Two	20	32	7	9	24	24	12	**128**	**42.1**
(f) Three	13	15	9	5	5	6	3	**56**	**18.4**
(g) More than 3	20	9	15	5	6	7	2	**64**	**21.1**

Source: Field Survey.

implementing the cooked MDMS in all eligible schools. As per State government reports, cooked meal is being provided to all children in government lower primary schools, government aided lower primary and Education Guarantee School (EGS) centres for 210 school days in a year. Assessment of the scheme carried out by District Institutes of Education and Training (DIET) of the State government in 2006-07 point to glaring lacuna on the functioning of the scheme in the State (GoM, 2007c).

Regularity of Meals and Coverage

Under the MDMS all eligible primary schools in the state[3] are to prepare and serve cooked meals to students every school day, which works out to 210 school days in an academic year usually starting from the middle of February to end of November.

However, the surveys found out that majority of sample schools were not serving cooked meal every school day. Only about one in four schools were serving cooked meals every school day, while 44 per cent were serving it on alternate school days. (Information gathered from the students also indicates that only about 23 per cent of schools were serving cooked meals every school day, while 35 per cent served it only on alternative school days). In the remaining schools meals were being served either once or twice a week or even once a month (Table 6.5). Further, in schools without a proper kitchen shed, meals are not cooked during the rainy seasons and instead uncooked rice is distributed to the students.

Although the scheme has a mandate to cover only children studying in primary section (till 2008), schools in the State are serving meals to children studying in other classes as well, particularly children in pre-primary section. The survey found that almost 92 per cent of the sample schools were serving cooked meals to children studying in pre-primary classes along with those in primary classes. In five other schools, meals were being served to children right from pre-primary to class seven (Table 6.5). It is no wonder that most schools are not able to serve meals regularly as the provisions meant only for students of primary level cannot last beyond few schools day when served to the whole school.

Meal Timings and Place

Under the MDMS, school meals are served in the day time in order to ensure that children who come to school empty stomach do not remain hungry throughout the session and also to ensure afternoon session attendance of children. However, the timing of school meal in the State is different from what is practised in the rest of the country. Most morning shift schools in the State (73%) are serving meals to children after the school hours generally between 9 am to 10 am. Even with day shift schools, many are found to be serving food in the morning hours before the start of the class between 9 am to 10 am.

Table 6.5: Regularity in serving Meals and Coverage

Districts/State	East Khasi Hills	West Khasi Hills	Jaintia Hills	Ri Bhoi	West Garo Hills	East Garo Hills	South Garo Hills	State	P—as percentage of State
(Number of schools)	58	65	34	20	62	44	21	**304**	**100**
As reported by schools									
(a) Every school day	7	6	6	4	38	10	3	**74**	**24.3**
(b) Alternate school day	27	33	4	12	15	25	18	**134**	**44.1**
(c) Twice a week	9	8	5	–	1	3	–	**26**	**8.6**
(d) Once a week	13	13	6	1	2	6	–	**41**	**13.5**
(e) Once/twice or thrice a month	2	5	13	3	6	–	–	**29**	**9.5**
As reported by students									
(a) Every school day	2	10	7	6	29	11	4	**69**	**22.7**
(b) Alternate school day	20	20	4	11	16	19	17	**107**	**35.2**
(c) Twice a week	8	8	5	–	4	3	–	**28**	**9.2**
(d) Once a week	17	12	7	2	2	9	–	**49**	**16.1**
(e) Once a month/ not regular	10	16	11	1	11	2	–	**51**	**18.8**
Classes provided cooked meals									
(a) Primary only	2	8	–	–	9	1	–	**20**	**6.6**
(b) Pre-Primary to primary	53	57	33	19	53	43	21	**279**	**91.81**
(c) Pre-Primary to class 7	3	–	1	1	–	–	–	**5**	**1.6**

Source: Field Survey.

Only 33 per cent of the day shift schools were found to be serving meals at noon, as practiced in other parts of the country. If we take into consideration both the morning and day shift schools, we find that nearly three-fourth of the sample schools are serving meals to the children in the mid-morning hours between 9 am to 10 am. In both day and morning section schools, whenever meals are served during school hours the subsequent classes are generally cancelled. The survey also found many schools (about 36%) served the meals inside the classrooms (Table 6.6).

Table 6.6: Place where Meals are taken and Meal Timings

Districts/State	*East Khasi Hills*	*West Khasi Hills*	*Jaintia Hills*	*Ri Bhoi*	*West Garo Hills*	*East Garo Hills*	*South Garo Hills*	*State*	*P—As percentage of State*
Number of schools	58	65	34	20	62	44	21	**304**	**100**
Meals are taken									
(a) Inside the classroom	20	28	25	13	9	2	12	**109**	**35.9**
(b) Outside	26	15	4	6	46	27	9	**133**	**43.8**
(c) In temporary shed	–	1	1	–	–	–	–	**2**	**0.7**
(d) In school hall	–	1	2	–	–	–	–	**3**	**1.0**
(e) Not specific	12	20	2	1	7	15	–	**57**	**18.8**

Meal timings(For the whole State—in percentage)

Morning section		**Day section**	
(a) At noon (12 to 1 pm)	15	(a) At noon (12 to 1 pm)	33
(b) After class	73	(b) After class	19
(c) Before class	5	(c) Before class	33
(d) Not specific	7	(d) Not specific	14

Source: Field survey.

Teachers in most of the sample schools were found to be actively involved in cooking and serving of meals as only about half of the schools had engaged a cook or helper for preparing the meal. Schools that had engaged cook/helper were found to selling part of the grain allotted to them in the market in order to raise the money to pay the cook/helper, while in few schools, students contributed towards paying the cook/helper. In some cases, cook/helper was given some quantity of grain in lieu of their service.

Infrastructure Issues

At the time of the survey, only 62 schools, which is 20 per cent of sample schools, had a constructed kitchen shed and in about the same percentage the kitchen was under construction. Thus, in 242 schools (80% of the sample schools) there is no proper kitchen shed and the food is being cooked either in the open or in nearby house or in temporary shed (see Table 6.7).

Table 6.7: Availability of Kitchen Shed

Districts/State	East Khasi Hills	West Khasi Hills	Jaintia Hills	Ri Bhoi	West Garo Hills	East Garo Hills	South Garo Hills	State	P—As percentage of State
Number of schools	58	65	34	20	62	44	21	**304**	**100**
(a) With kitchen shed	5	7	2	1	21	11	15	**62**	**20.4**
(b) With kitchen shed under construction	10	11	12	3	13	11	–	**60**	**19.7**
(c) Without kitchen shed	43	47	20	16	28	22	6	**182**	**59.9**
Schools without kitchen (b + c)	53	58	32	19	41	33	6	**242**	**79.6**
Place where food is cooked									
(a) Open air in the school compound	36	22	25	10	32	25	5	**155**	**64.1**
(b) Nearby house	8	31	2	3	6	4	1	**55**	**22.7**
(c) Temporary shed	9	5	5	6	3	4	–	**33**	**13.2**

Source: Field survey.

In Meghalaya, the rainy seasons usually starts by the middle of May and continues right to the end of September[4]. Therefore, in schools without proper kitchen shed, food is being cooked only in the first-three months of the year (mid February to mid of May) and in the last two months of year (October and November) and that only when the foodgrains and cooking costs are issued to schools. Thus, in the absence of a kitchen shed, in many schools cooking of meals ceases during the rainy seasons. This is one of the main impediments in the proper implementation of the scheme in the State.

Along with kitchen shed, the availability of storeroom is an important requirement as most of the sample schools are either one or two room schools with hardly any space for storing of like rice, pulses, vegetables and cooking ingredients. In the absence of a kitchen-cum-storeroom in most of the sample schools, provisions are being kept in places away from the schools such as house of teachers/village headman/others (31%), which raises the possibility of misuse of the provisions.

While almost all of the sample schools had the necessary utensils for cooking purposes, only nine schools have containers to store the provisions (see Table 6.8) Children in most of the schools bring their own plates (in some case leaves) from home whenever meals are served or else schools arrange for leaves in which the meals are served to the students.

Table 6.8: Availability of Utensils, Containers and Storeroom

Districts/State	*East Khasi Hills*	*West Khasi Hills*	*Jaintia Hills*	*Ri Bhoi*	*West Garo Hills*	*East Garo Hills*	*South Garo Hills*	*State**	*P—As percentage of State*
Number of schools	58	65	34	20	62	44	21	**304**	**100**
Schools where utensils are									
(a) Available	58	64	30	20	57	42	18	289	95.1
(b) Not available	–	1	4	-	5	2	3	15	4.6
Schools where containers are									
(a) Available	4	-	1	2	2	–	–	9	3.0
(b) Not available	54	65	33	18	60	44	21	295	97.0
Provisions are kept in									
(a) Classroom	18	6	9	1	8	7	2	51	16.8
(b) School office/head master' s room	13	17	7	16	5	4	8	70	23.0
(c) Kitchen/separate storeroom	8	10	2	2	25	14	11	72	23.7
(d) Teachers/secretary's house	11	28	14	1	21	18	–	93	30.6
(e) Nearby house/ others	8	4	2	–	3	1	–	18	5.9

* Schools in South Garo Hills district were not included in this analysis as many of them did not provide definite answers on this issue.

Source: Field survey.

Only in 27 per cent of the sample schools there is a source of water inside the school premise, while in another 10 per cent there is filtered drinking water available to the students. In remaining schools, water that is used for cooking is brought in containers and buckets from nearby source for which school children and in some cases even the parents are involved in fetching water.

Concerning Logistic Foodgrains and Cooking Costs

One of the main reasons adversely affecting the proper implementation of the scheme by schools in the State is the irregular supply of grain to schools. For example, during the time of the survey in the month of May-June of 2008, sample schools had received only one allotment of rice for the period of February to April for that year, and that too was released only in the month of April. According to office of DIs, the erratic release of foodgrains from their end is mainly because of the delay in issuing of lifting report by the department of Food and Civil Supply of the State government without which grain cannot be allotted to schools.

On the question of the whether the schools get the full quantity of grain allotted to them, almost all schools (about 89%) indicated that they do get the full quantity allotted to them. However, a common complaint from some school is that the dealers usually reduce some quantity of the foodgrains on the pretext of not having the required unit to weigh small quantities. Around half of the sample schools described the grain issued to them as being of good quality, while around 45 per cent rated them to be of fair quality. Some schools (6%) reported of not getting good quality of foodgrains which is returned to the dealer and exchange for better quality. However, for schools located in far flung places this is not possible because of the huge expenditure involved in transportation foodgrains from the dealer to the schools. In such schools the bad quality grain is simply distributed to the students.

As per the scheme, foodgrains is to be delivered by the dealer to the schools free of cost for which the dealers are reimbursed by government. However, the findings of the survey points to a totally different system prevailing in which majority of the sample schools have to bear the cost of lifting and transporting the foodgrains from the dealers to the schools. Only 17 per cent of the schools surveyed reported that foodgrains was being delivered to the schools free of cost, while in four schools the villagers or the students themselves carried the food grains from the dealer's shop located near the school. In all other cases (about 82%), schools have to bear the cost of transportation of foodgrains. This includes 39 schools where dealers delivered foodgrains to schools but charged the schools the cost of transporting. The survey found that the school had different arrangements for raising the amount spent on transportation of

Table 6.9: Arrangements for Delivery of Foodgrains to Schools

Districts/State	East Khasi Hills	West Khasi Hills	Jaintia Hills	Ri Bhoi	West Garo Hills	East Garo Hills	State	P—As percentage of State
Number of schools	**58**	**65**	**34**	**20**	**62**	**44**	**283**	**100**
Arrangements for supply of foodgrains to school								
(a) Foodgrains delivered by dealer free of cost	0	0	30	–	13	4	47	16.6
(b) Arrangements not involving transport cost transportation cost	1	0	–	–	3	–	4	1.4
(c) Schools bear transportation cost	57	65	4	20	46	40	232	82.0
Expenditure on transportation is met by/from								
(a) Cooking cost	9	22	–	2	18	19	**70**	**30.2**
(b) Cooking cost and contribution from managing committee (MC)/school funds/ teachers	11	3	–	–	–	–	**14**	**5.0**
(c) Cooking cost and selling of foodgrains	1	6	–	–	–	2	**9**	**3.9**
(d) Selling of food grains	6	1	1	–	6	10	**24**	**10.3**
(e) Selling of food grains and contribution from teachers/ MC/ school funds	2	1	–	–	–	–	**3**	**1.3**
(f) Contribution from MC/ school funds/teachers	18	31	2	16	14	8	**89**	**34.3**
(g) Contribution from children	7	–	1	–	–	–	**8**	**3.4**
(h) Other arrangements (not specific)	3	1	–	2	8	1	**15**	**6.4**

Source: Field survey.

foodgrains which included using part of the money meant for cooking costs and also by selling some quantity of the foodgrains in the market. In other schools, the transportation amount was raised by contribution from school management, teachers and in some cases even the schools children themselves (see Table 6.9).

Aside from the problem concerning the delivery of foodgrains, the issue of delay in the release of cooking costs is another main complaint of the schools regarding the MDMS in the State. Information supplied by the office of the DIs on cooking costs released to schools included in the survey in 2007 and 2008 (August) reveals the following:

1. In 2007, cooking costs was released to schools in the month of July to August 2007 which was for 86 school days covering the period **June to October 2007**.
2. The cooking costs for remaining school days of upto the month of May 2008 was released only in **April to May 2008,** which was for 116 days covering period **November 2007 to May 2008.**
3. Cooking costs for the remaining schools days was yet to be released as in the time of the survey (mid 2008).

Given this situation in which foodgrains and cooking costs are not released to the schools in time, it is only those schools that are able to mobilize the provisions and ingredients for the meals, either by using or borrowing money from the school funds or by buying the ingredients on credit, that are able to implement the scheme regularly and as per norms.

Perceptions of Teachers

Despite the many problems faced by schools in implementing the scheme as discussed above, there is widespread support for the scheme from teachers. Many teachers support the scheme as they felt that if properly implemented it would boost school enrolment and also raised the daily attendance of children. According to them, even in the present situation, servings of meals by schools have managed to attract children to schools. However, at the same time, teachers in many of the surveyed schools have also pointed out that the scheme does disrupt classrooms teaching as they have to help and supervise in the preparation and serving the meals. The absence of proper facilities such as kitchenshed-cum-storeroom in schools also has an adverse impact on the teaching process.

Concluding Remarks

An interesting point of interest thrown upon the survey is with regard to meal timings. Unlike the common practice followed by schools in the rest of the country of providing meals to students in the noon time, schools in the State are found to be serving the meals during the mid-morning hours between 9 am to 10 am and this include a significant percentage of day shift schools. While for day shift schools there is no reason to support serving of meals in any other timing other than in the noon; the same cannot be applied to morning shift schools. Given that many of the primary schools in the State particularly those in the rural areas are morning section schools, there is a need to find out the appropriate time for providing school meals in such schools keeping in view the eating habits of such children in order to ensure that the objectives of elimination of classroom hunger and achieving better educational outcomes are realized.

On the whole, we find that the functioning of the MDM scheme in Meghalaya leaves much to be desired. The findings of the survey clearly shows that the implementation of the MDMS in the Meghalaya is plagued by many problems such as irregular flow of funds and foodgrains to schools, clear deviations of guidelines of the scheme and absence of supporting infrastructure in the schools. A large part of the blame for the shoddy implementation of the scheme in the State lies squarely with the State government. Delay in release of funds and foodgrains to the schools is one of the major reasons preventing schools in providing cooked meals to schools on regular basis. The existing system where schools have to arrange and pay for the transportation of foodgrains is a clear violation of the guidelines of the scheme and it compels schools to misuse the funds and foodgrains allotted under the scheme in order to raise the amount spent on transportation. The lack of facilities such as kitchen in schools also adds to the challenge of ensuring regular cooking of meals particularly during rainy seasons. Given the importance of MDMS in improving the enrolment, attendance and retention of school children and the low nutritional status of children in the State, it is imperative that the MDMS is properly implemented to improve the performance of the primary education and realize the goal of UEE in Meghalaya.

NOTES

1. Evidence of the high incidence of malnutrition among children in the State is also supported by field studies undertaken by Agrahar (2005) and Khongsdier & Mukherjee (2003).
2. Of the 15 DIs in the State only five responded to questionnaire sent to them.
3. As per revised guidelines for MDMS issued in 2006 for North-Eastern States like Meghalaya the Central government provides each child at the primary level with 100 grams of foodgrains and cooking costs at Rs. 1.80 (provided the State government contributes a minimum of 20 Paise) per school day. The Central Government also provides for reimbursement of the actual cost incurred in transportation of foodgrains from nearest FCI godown to the primary school.
4. Some of the places in the State like Mawsynram and Cherrapunjee are wettest place on earth with an annual rainfall of around 11.500 metres.

REFERENCES

Afridi, F. (2005). Midday meals in two States: Comparing the financial and institutional organisation of the programme. *Economic and Political Weekly*, 40(15): 1528-35.

Agrahar, M. D. (2005). Nutritional Status of Khasi school girls in Meghalaya, *Nutrition*, 21(4): 425-431.

Dreze, J., and Goyal, A. (2003). Future of mid-day meals. *Economic and Political Weekly*, 38(44): 4673-83.

Dreze J., and Kingdon, G. G. (2001). School participation in rural India. *Review of Development Economics*, 5(i):1-24.

GoI (2001). *Census of India: Primary Census Abstract*. New Delhi: Office of the Registrar General, Government of India.

____ (2008). *Eleventh Five Year Plan 2007-12*. New Delhi: Oxford University Press.

GoM (2007a). *Statistical Hand Book of Meghalaya*. Meghalaya: Directorate of Economic and Statistics, Government of Meghalaya.

_____ (2007b). *District Level Statistics*. Meghalaya: Directorate of Economic and Statistics, Government of Meghalaya.

______ (2007c). *Quarterly assessment of the implementation of Mid-Day Meal Scheme in districts of Meghalaya, District Institute of Education and Training (DIET) (South Garo Hills, East Garo Hills, West Garo Hills, West Khasi Hills, Jaintia Hills, Ri Bhoi)*. Meghalaya: Government of Meghalaya.

_______(2009). *Annual Plans 2009-10*. Meghalaya: Government of Meghalaya.

Jain, J., and Shah M. (2005). Antyodaya Anna Yojana and Mid-Day Meals in MP. *Economic and Political Weekly*, 40 (48): 5076-88.

Khera, R. (2006): Mid Day meals in primary schools: Achievements and challenges. *Economic and Political Weekly*, 41(46): 4742-4750.

Khongsdier, R & Mukherjee, N. (2003): Growth and nutritional status of Khasi boys in North-East India relating to exogamous marriages and socioeconomic classes, *Am. J. of Physical Anthropology,* 122(2): 162-170.

Mehta A. C. (2011). *Elementary Education in India: Progress towards UEE.* New Delhi: National University of Educational Planning and Administration.

MHRD (2009). *Annual Report (various years)*. Ministry of Human Resource Development. New Delhi: Government of India.

RHM (2007). *Rural Health Statistics*. New Delhi: Government of India.

Samson, M; Noronha, C and De, A. (2005). *Towards more benefits from Delhi's Mid-Day Meal Scheme: Collaborative Research and Dissemination*, New Delhi. Retrieved from *http://www.cordindia.com/images/Midday.pdf* (accessed January 15, 2010)

Sinha, D. (2008). Social audit of midday meal scheme in AP. *Economic and Political Weekly*, 43(44): 57-61.

Umdor, S. and Panda B. (2009). State of infrastructure in Meghalaya, in *Meghalaya State Human Development Report 2008*. Meghalaya: Government of Meghalaya.

7

The Economic Costs of Insurgency: Some Explorations in Manipur

K. Gyanendra Singh

There are two-way linkages between economic underdevelopment and insurgency. An underdeveloped material base provides little opportunities to people for gainful employment. Unemployed youths then hardly find any viable alternative for their livelihood. They perceive that the 'failed State' is the prime cause for their misery. This perception encourages them to join the non-State actors who vouch for revolutionary change. Thus economic underdevelopment helps in sharpening anti-State political discourse and strengthens the ideology of the so called revolutionary groups.

Insurgent activities help in perpetuating economic underdevelopment. Militancy has created deep insecurity in the minds of the people in general and entrepreneurial class in particular. Majority of the entrepreneurs in the State hail from outside—be it large business in goods or services. These 'others' are really vulnerable as they are often considered as outcast and pejoratively known as *mayang* or 'outsider'. They often become the soft targets and easy source for extortion. The economic space held, hitherto, by these *mayang* is gradually being restored to the sons-of-the-soil through a strong nexus among the politicians, local contractors and insurgents. The *nouveau riche* Meetei entrepreneurial class who emerged out of State patronage following the coming of the State of Manipur into existence since 1972 appears to have been acting as a cog in the wheel of the militancy machine. They appear to play the role of the interface between the power elites and the undergrounds particularly in siphoning off development funds earmarked for public works. Thus while private investments shy away with the rising risk factor, public investment is squandered away in the name of development leading to the perpetuation of economic underdevelopment making the socio-political space fertile for the growth of insurgency in Manipur. A Nurkesian type of vicious cycle between underdevelopment and insurgency has trapped the economy of the State at low level equilibrium which calls for an appropriate policy intervention. However, the economic costs of insurgency manifested in different forms are not easy to quantify. It is all the more difficult to establish empirical association between them. This, however, should not deter us to look into the economic costs of insurgency in terms of descriptive and consequential analysis.

Economics of Insurgency: Crippling Effects of Rent Seeking (Extortion)

The militancy machine requires huge resource mobilization to wage war against the State. Besides the maintenance of the cadres and camps, they need huge amount of cash for procuring arms and ammunition primarily from the arms markets abroad, arranging safe shelters for the leaders and cadres outside the State and abroad, arranging training for the cadres in hideouts of other insurgent groups, undertaking subversive operations, replenish the men and materials lost in operations and for other logistics. Unlike the civil war groups in African countries, Manipur does not have any natural resources of any significance on which the militant groups can fall back on for resource mobilization. Neither the State has large business operators to be milked, nor the communities have strong diasporas to bank on. As a result, the lion's share of the resources is primarily mobilized from within.

Moreover, as the constituency of the ethnic militant groups is primarily limited by the ethnic boundary, it is difficult for one ethnic militant group to compel people of other ethnic groups to pay for their struggle. As for illustration, a Meetei militant group keeps away from the Nagas or Kukis, and a Naga militant group keeps away from the Meeteis and Kukis. As the different ethnic militant groups often fight for overlapping interests, allowing the 'other ethnic-militant groups' to operate within the domain of 'us-militant-group' will only rob the loyalty and weaken the social base of the latter. As a result, ethnic boundaries are jealously guarded by the respective ethnic militant groups. Of course, for operational purposes, homeland concept is generally used for resource mobilization. As ethnic boundaries and homeland boundaries, by and large, coincide, people in the valley pay to the Meetei insurgent groups, whereas people in Naga inhabited hills pay to Naga insurgent groups and people living in the Kuki inhabited areas fill the coffer of the Kuki insurgent groups. However, in isolated pockets and ethnic enclaves, the scenario may be different. Thus, the resource domains of the insurgent groups are largely limited by the territorial boundaries of the homelands of the different ethnic groups.

Furthermore, the existence of multiple militant groups within a particular ethnic community further reduces the resource domain of a group and leads to conflicting loyalty claims. The proliferation of ethnic armies and inter-group competition imposes heavy demand on the resources of the people. Today Manipur tops the Union Government's list of underground groups with seven valley based groups being declared unlawful under the sub-section (1), section 3 of the Unlawful Activities (Prevention) Act 1967 (Ministry of Home Affairs, 2011-12).

The instruments of resource mobilization adopted by the majority of the groups constitute extortion or often eulogize as 'protection tax' or 'monthly donation for freedom struggle', ransom or captive tax primarily collected through kidnapping, commission for public works and development projects, fines from the wrong doers collected from the corrupt officials, and 'beneficiary tax' mainly collected from the contractors and suppliers who act as the surrogate of the militant groups.

In absence of any alternative source for resource mobilization, the burden of extortion largely falls on the people in general and salaried, professional and business groups in particular. The multiplicity of extortion demands due to the existence of multiple groups has made it very difficult for the people to comply with. The level of extortion has indeed surpassed the level of tolerance. Informal discussions that we had with different people from different walks of life suggest the current extent of extortion to the tune of 10-12 per cent of the net pay of the salaried groups. What is more important is the question of compliance. It has been learnt that the dreaded the image of a group, the higher is the level of compliance. In order to have a dreaded image, different groups adopt different code of conduct for non-compliance. Some opt for kidnapping, some for shooting at the leg, some for outright killing, and some for creating panic by throwing hand grenades. Unlike the militant groups in Assam who extort bigger amount mainly from the limited number of large players, Meetei militant groups deal with innumerable small players to make up a large amount. Hence, they have to face a difficult task as far as the compliance is concerned.

Except the Army and the Central Paramilitary Forces who are engaged in counter insurgency activities (Routray, 2007), the wide extortion network has affected every earning citizen of the State. All the groups resort to extortion, kidnapping, murder, etc. The extortion racket of the underground groups works in a well channelised path. Almost all the active groups used to collect 'monthly donations' from thousands of government employees and other salaried groups. Apart from these they collect money from all the business houses including even the petty shopkeepers on regular basis. All the shopkeepers have to pay a sum regularly to the underground groups. This sum ranges between Rs. 300 to 500 to 1000 and even more per annum depending on the size of the business and negotiations. In their extortion racket they have not even spared petty shopkeepers and rickshaw pullers.

In January 2004, police shot an 'unidentified suspect' who was found not only having a hand grenade but also receipts of taxes collected the previous month. These included amounts of Rs. 10,000 from Education Department, Rs. 12,000 from the Public Health Department, Rs. 5,000 from minor irrigation projects, Rs. 7,000 from the Medical Department and Rs. 15,000 from the Public Works Department (Parratt, 2005: 170).

Apart from monthly and annual donations, the underground groups often resort to special drive for collection. They used to demand a huge sum amounting to lakhs of rupees from the well-to-do section as one time donation ostensively to meet the emergency need of the organizations.

Besides cash, sometimes tax is also collected in kinds. The militants ask for supplies like jackets, shoes, clothes, medicines, mobile sets, motorcycles, fuels, etc. In villages, although they do not demand cash, but tax in kinds like eatables, shelter and hospitality are demanded.

As far as the system of collection of extorted fund is concerned, the militant groups follow different methods for different groups of people. For salaried class of people, they follow exactly the pattern laid out by the income tax authority in the country. They collect the booty from the disbursement desk in one shot for all the employees in an office. The 'monthly donation' of the employees is deducted at source based on the agreed upon formula which ranges from 1.5 to 3 per cent depending upon the strength of a group. Thus public offices have been tamed to work as the back office of the insurgent groups. In order to buy peace, most of the public offices and institutions persuade their employees to comply with the dictates of the militant groups. As the State has utterly failed to protect the life and property of its citizens, denial to comply only invites death. In case of business, most of the groups settle for annual donation. Unlike the salaried group, there is no mechanism for collective payment. The militants, based on the perceived business turnover, negotiate an annual amount and get them collected either through agents or cadres.

As has already been mentioned, it is not only the amount of extortion which is a fear factor but the actions of the underground groups on the failure to accede to their demands. Evidences are plenty when the underground groups have resorted to severe actions when their demands are not met. In a press release by Kangleipak Communist Party (KCP) on September 20, 2004, justified its extortion activities and said "the revolutionary movement could survive only on the goodwill of the people". It asked the people to honour the 'demand notes' and also warned of punitive actions in case anybody 'dares to defy'. On April 25, 2005, Kangla Yawol Kanna Lup (KYKL) announced its 'new tax policy' where it was stated that the outfit will "stop serving donation letters for huge amounts and will collect extensive tax from the people". It was further stated that "money in bulk will continue to be collected from the corrupt officials, outsiders and from those who earn their money through dubious means" (www.satp.org). In 2002 KYKL (Oken) shot Dr. Bedikumar of Regional Institute of Medical Sciences (RIMS) in leg because, it alleged, he had refused to levy its tax of 1.5 per cent on RIMS employees. Again in 2004 the same group shot an officer of the State's Horticulture and Conservation Department in the thigh in pursuance of their demand for no less than Rs. 57 lakhs from the department (Parratt, 2005: 170-171). Another incident which shocked everyone was the gunning down of three Catholic priests of the Don Bosco Training Centre at Ngariyan hill, about 10 km south-east of Imphal on May 15, 2001. The police version was that KYKL (Toijamba) was behind the act for not paying the full amount of the huge monetary demand but the outfit denied the charges (Tarapot, 2005: 280). Again on November 10, 2005, KYKL exploded a bomb in Imphal city targeting a particular shop for refusing to pay the demand of tax. On October 21, 2005, People's Revolutionary Party of Kangleipak (PREPAK) attacked the residence of Arambam Binodini, Joint Director, Department of Tribal Development and Other Backward Classes for failure to meet the deadline for payment of extortion money. On November 27, 2006, the same group locked up main office of Manipur State Agriculture Department at Sanjengthong for refusal to pay money and again on December 12 of the same year the outfit hurled a bomb at the residence of the Director of Agriculture (www.satp.org).

These are only the tip of the iceberg of the various incidents relating to extortion and consequences of non-compliance. There are many more such incidents. Several senior officials including Chief Engineer (PWD), S. Binoykumar Singh; Chief Engineer (PHE), Thingom Ibochouba, Loktak Hydro Electric Project; Chief Engineer, Subhas Chandra Ser, Director (Health services); Surchand Singh, a telecom department engineer, Ambe, Director (Tourism); Dasarath Prasad (IAS), executive engineer of Singda Dam project; A K Mathur, an assistant accounts officer of the project; T. K. Das, a contractor of the project; Phoni Bushan Verma, etc.; fell to the bullets of the different underground groups for refusing to concede to their pressures (Tarapot, 2005: 279-280).

Highway extortion is another important source particularly for the groups operating in the hills through which the lifelines of Manipur run. The Naga and Kuki groups collect 'taxes' from all the vehicles plying through both the Imphal-Dimapur-Guwahati and Imphal-Jiribam-Silchar national highways since 1990. On August 6, 1996, the All Manipur Road Transport Drivers and Motor Workers' Union (AMRTDMWU) and Manipur Truck Owners Welfare Association submitted a memorandum to the Union Home Minister stating that the National Socialist Council of Nagaland (Isak-Muivah) [NSCN (I-M)] cadres in the name of Government of People's Republic of Nagaland (GPRN) were collecting 'various taxes' ranging between Rs. 1,000 to Rs. 10,000 from every vehicle plying along these two highways. In another memorandum on March 12, 1999, the AMRTDMWU sought protection of road transporters and passengers on the highways alleging that the NSCN (I-M) members 'demanding and extorting with receipts in the name of GPRN 'from each and every truck as goods tax in the following rates: (1) from every gas cylinder truck, Rs. 2,000 per trip, (2) from every oil tanker, Rs. 3,000 per trip, (3) all the trucks carrying different goods: (a) truck carrying iron rods and scrap irons, Rs. 3,000 per trip; (b) trucks carrying cement, Rs. 1,000 per trip; (c) trucks carrying rice from FCI godown, Rs. 500 per trip; (d) trucks carrying essential commodities such as rice, pulses, edible oils and others, Rs. 700 to 800 per trip; (e) truck carrying sand, Rs. 300 per trip. The memorandum also pointed out that the NSCN (I-M) collected road taxes from the vehicles on the basis of Rs. 12,000 from every tourist bus, Rs. 7,000 from every truck and Rs. 10,000 per annum per agent of gas cylinder (Tarapot, 2005: 57-58).

The most important source of rent seeking by the militant groups constitutes public works, development schemes and development projects. It is learnt that all the groups together collect 10-12 per cent of the project allocation from the implementing authorities and the booty is then divided among the groups based on certain agreed arrangement. This sum is collected through the public agencies at the district level. Very often, development projects are routed through a class of contractors either drawn from the close aids of the insurgent groups or from their commission agents (Das, 2002). This intense rent seeking activities of the militant groups have a number of ramifications for the economy of the State.

First, insurgency legitimizes corruption. As the people working in public domain have to pay to the insurgents, they develop a tendency to get it compensated through corruption. Since the militant groups stand solidly behind these people, they get emboldened in their pursuit. Politicians, contractors, business class—all follow the footprints of the insurgent in sucking the development funds meant for the welfare of the ordinary citizens particularly the poor. It is learnt from discussion with the informed group of people that rent seeking by the people who control the State power is no less a menace than their insurgent counterparts. It is believed that around 45-50 per cent of a development project or public work is being siphoned off by both the State and non-State actors put together. The only difference is that while a part of the ill-earned money is ploughed back in the local economy by the State actors, the lion's share of the booty in the hands of the militant groups finds its way out of the State and very often out of the country to buy arms and other logistics. A conservative guesswork suggest that multilayer modes of rent seeking by the militants involves at least 15-20 per cent of the budgetary allocation of the State which stands to the tune of Rs. 633 crores as per the 2007-08 budgetary expenditure of the State. Thus, at least one-fifth of the budget of the State goes to oil the militancy machine of the State. Assuming half of this amount going for counter-insurgency operations, rent seeking by the State actors and the intermediary agents like the contractors and suppliers, another Rs. 317 crores is withdrawn from the productive services. Thus the actual size of the economy gets reduced to one-third due to the combined rent seeking from all quarters. This has a pernicious impact on the growth and development of the State.

Second, rent seeking by the insurgents acts as disincentive against entrepreneurial initiatives and capital formation particularly in non-farm urban sector, as the farm sector is not heavily targeted, of the economy. It penalizes the entrepreneurs for undertaking wealth and income creating pursuits. Any signs of accumulation, asset formation and wealth creation are immediately greeted with extortion notes. As a result, very often the idea dies before it fructifies and the baby is thrown away with the bath water. Only the entrepreneurships of violence thrive only to extend the horizon of despair and destitution for the ordinary citizen in the State.

Third, rent seeking behaviour of the insurgent groups and their collaborators leads to the low quality of public works. As has already been mentioned, 45-50 per cent of all project allocation, in general, is leaked in the pipeline and hence the quality of the project suffer and the high maintenance cost that follow make most of them economically unviable. This causes poor quality of infrastructure which in turn retards economic and social development.

Fourth, besides the actual economy, the rent seeking activities of the militant groups have tremendous effects on the potential economy. Insurgency directly affects business confidence and investment climates. Due to the high risk factor, private investment is hard to come by. Rather the business enterprises owned particularly by the '*mayangs*' including the Marwaris, Punjabis, Sindhis and Bengalis are investing a part of their profit outside Manipur as a long term strategy since if they are to wind up their business in the State they may fall back upon the alternative business opening elsewhere. This is leading to a large scale flight of capital from the State through non-institutional channels. Thus insurgent activities have shut the doors of investment only to cause the economy to languish in the morass of underdevelopment.

Flight of Institutional Capital

It has already been mentioned in the previous section that deteriorating security environment has led to flight of capital through both institutional and non-institutional channels. Besides the business sector and insurgent groups, a significant amount of fund also goes out of the State by way of transfer income. A sizeable segment of the student population from the State migrate every year to other parts of the country particularly to places like Delhi, Bangalore, Pune, Shillong and Guwahati for acquiring formal education. As the parents want to keep their wards from the war theatre, those who can support send their children out of the region. Safety of the child is, in fact, paramount in the minds of the parents. This causes a huge transfer of income from the State which is required for the upkeeping of the children. Moreover, as Manipur is neither self-sufficient in food production nor has any industrial base of its own, a huge amount of fund moves out of the State in order to procure goods and services. It is, thus, not possible even to arrive at any reasonable conjecture as to what is the size of the transfer of fund through the non-institutional channel.

The flight of capital through the institutional mechanism can be partially captured through the trend in credit-deposit ratio of the scheduled commercial banks in the State.

It may be noted from Table 7.1 that the number of bank branches has increased from 84 in 1992 to 87 in 2000. But this expansion of banking area could not be sustained primarily due to the problem of security. The number has gone down to 77 in 2004. Ten branches in the interior areas had to be closed down due to poor security environment. As far as the C-D ratio is concerned, it has declined from 84 per cent in 1993 to 25 per cent in 2002, a whopping 59 percentage point fall in 10 years. This fact indicates the tumultuous condition through which the banking system in the State had passed. A complete loss of confidence either on the part of the banking system on the prospect of loan recovery or the low business prospect perceived by the entrepreneurial class might have contributed to this great fall. After touching the bottom line in 2002, C-D ratio for the state started looking up and reached at 53 per cent in 2007.

Table 7.1: Deposits and Credit of Scheduled Commercial Banks of Manipur (as on 31st March)

Year	No. of Banks	Deposit (Rs. In Lakhs)	Credit (Rs. In Lakhs)	Credit-Deposit Ratio (per cent)
1992	84	10916	7925	72.60
1993	85	10511	8865	84.34
1994	85	14576	9524	65.34
1995	85	17742	9865	55.60
1996	85	23093	12406	53.72
1997	85	26421	15195	57.51
1998	85	28343	16657	58.77
1999	86	40622	16923	41.66
2000	87	46100	16800	36.44
2001	80	42300	16500	39.01
2002	79	61100	15500	25.37
2003	79	63900	18000	28.17
2004	77	81900	23800	29.06
2005	78	97200	39800	40.95
2006	78	121400	61500	50.66
2007	77	144000	76900	53.40

Source: Economic Survey Manipur 2006-07 and 2007-08.

Table 7.2: Credit-Deposit Ratio of Scheduled Commercial Banks in the North-Eastern States (as on 31st March)

State	1996	1998	2001	2002	2004	2005	2006	2007
Arunachal Pradesh	10.4	13.1	17.2	15.8	17.2	24.4	25.07	26.81
Assam	39.3	32.9	32.4	31.4	30.8	34.4	41.97	43.30
Manipur	53.7	58.8	39.0	25.4	29.1	41.0	50.66	53.40
Meghalaya	14.4	15.2	16.8	18.0	36.9	45.3	39.36	35.84
Mizoram	16.2	23.2	25.6	25.9	38.3	51.0	51.52	53.89
Nagaland	27.4	18.3	14.0	12.6	16.9	22.9	22.32	28.90
Tripura	42.1	34.0	23.0	22.3	25.4	29.7	31.63	33.98
All India	61.9	55.5	58.5	62.3	58.7	66.0	72.50	75.02

Source: Basic Statistics of NER 2006, *Economic Survey Manipur 2002-03, 2006-07 and 2007-08.*

However, the C-D ratio of Manipur rather looks moderately good while compared with the same in other North Eastern States (Table 7.2). In fact, Manipur is only marginally trailing behind Mizoram while the other North-Eastern States are lagging far behind. One probable reason for a relatively higher C-D ratio in Manipur might be traced to the system of determination of extortion amount from many small players by the insurgent groups of Manipur. The extortion/donation is based on the net pay of the salaried segment of the population as has already been pointed out. This fact encourages the salaried groups to borrow from the banks for buying the durable consumer goods, house sites, house construction, etc. As the repayment is made through equal monthly instalment which is deducted from the monthly salary, the quantum of net salary gets reduced and hence one has to pay less to the coffer of the insurgent groups. This hypothesis, however, needs to be verified. No studies have yet been undertaken to find the impact of the extortion demand on the consumer behaviour. As a result future studies on this aspect might be revealing for the stakeholders of market economy.

Low Economic Growth

The decennial compound growth of Net State Domestic Product (NSDP) is declining over the years. It has declined from 7.37 per cent in 1961-70 to 6.84 per cent and 4.80 per cent in 1971-80 and 1981-90 respectively. It further declined to 4.68 per cent in 1991-2000 (Table 7.3).

Similar trend is also found in case of per capita income. The decennial compound growth rate of per capita income declined from 4.16 per cent in 1961-70 to 3.81 per cent and 2.14 per cent in 1971-80 and 1981-90 respectively (Table 7.3).

Table 7.3: Decennial Compound Growth Rate of NSDP and Per Capita Income in Manipur

Year	*NSDP*	*Per Capita Income*
1961-70	7.37	4.16
1971-80	6.84	3.81
1981-90	4.80	2.14
1991-2000	4.68	2.36
2001-07	11.30	9.06

Source: Calculated from EPW and CSO.

What is important to note is that the 7.37 per cent decennial compound growth rate of NSDP for the decade 1961-70, which may be called as pre-insurgency period, was quite high. In fact the growth rate of Manipur was more than double to that of the national average. It has come down slightly during the early years of insurgency in the State during the 1971-80. Even then growth rate of NSDP in Manipur remained far higher than the national average (3.23%). However, with North-Eastern region becoming the flashpoint of ethnic war during the 1980s particularly following the formation of NSCN (I-M) in Nagaland, ULFA in Assam, TNV in Tripura and PLA, KCP and PREPAK in Manipur, NSDP of Manipur has dipped down by more than two percentage points and fell below the national average (5.11%). The intensity of militant activities remained very high although the 1990s in spite of peace initiatives like Assam Accord in 1985 and Mizo Accord in 1986. With the outbreak of Naga-Kuki conflict in 1992 and emergence of KYKL, the security environment in Manipur deteriorated to all time low and the NSDP had also touched to its all time low. However, following the adoption of 'new initiative for North-East' in 1996 by the central government and subsequent creation of the central pool of resources for the development of North-East as well as ceasefire agreement between the NSCN (I-M) and Government of India in 1997 that the economy of Manipur started looking up during 2001-07.

Thus, although it is difficult to establish inverse statistical association between the NSDP and insurgency, nonetheless qualitative evidences suggest that insurgent activities have severely affected the growth of the economy of Manipur.

Low Tax Base and Economic Dependency

Low economic growth has a significant bearing on resource mobilizing capacity of the State. As the size of the wealth creating population is small, the tax base has become very small. Moreover, since a significant size of the working population belongs to the scheduled tribes, who are out of the purview of taxation, tax revenue forms only an insignificant proportion of the total budgetary expenses of the State of Manipur. The size of the non-tax revenue largely depends on the resource endowments as well as the entrepreneurial skill of the State manifested particularly in the success of the public sector, as Manipur is neither resource rich nor has succeeded in creating a vibrant public sector enterprises, the non-tax revenue is also awfully insignificant. As a result, the State has failed to attain economic viability. The obvious outcome is the rise in fiscal dependency on the centre. While economic logic calls for gradual reduction of the dependency, which, in turn, necessitates creation of a competitive business environment, insurgent activities keep on pulling the economy in opposite direction. As no nation affords to live long on doles without compromising with pride and relative autonomy, insurgents, by waging political war and undermining the economic strength of the State, are standing on the way of consolidation of the Manipuri nationalism.

In case of Manipur we find that the share of tax revenue to total revenue receipts is very low. All through the years since 1980s, the share remained below 5 per cent. The same is the case with non-tax revenue whose share remained below 5 per cent all through the years (Table 7.4). This has led to economic dependency and the State is surviving on funds from the centre. The share of grants-in-aid from the centre account for more than 70 per cent of the total revenue receipts of the State. It may be recalled that the State also enjoys a liberal funding, being a part of Special Category State (SCS), and receives 90 per cent plan assistance as grant and 10 per cent as loan, whereas States other than SCS get 30 per cent plan assistance as grant and remaining 70 per cent as loan.

Table 7.4 shows that a major part of the revenue comes from the centre. The State's own revenue which comprise of State's own tax revenue, State's own non-tax revenue and share of central taxes was 36.33 per cent in 1981. Since then it has declined to 34.76 per cent in 1990. It increased to 43.11 per cent due to increase in the share of central taxes in 1995 but it has declined to 37.43 per cent in 2000. Since 2000 the State's own revenue has declined due to decline in State non-tax revenue and share of central taxes. The State's own share in the total revenue receipts of the State is below 30 per cent all through the years since 2000. In 2001 State's own share was 24.34 per cent of the total revenue receipts of the State. Its share had increased to 25.25 per cent in 2004 which came down to 21.32 per cent in 2006. It has increased marginally to 22.99 per cent in 2007 and to 28.61 per cent in 2008. One interesting feature is that the share of non-tax revenue is continuously declining over the years. Only recently in 2007 and 2008 the share of non-tax revenue has increased to 5.52 per cent and 5.30 per cent respectively. Even the share of central taxes kept on fluctuating all through the years. With low tax revenue the State has become totally dependent on the centre for its functioning. The grants-in-aid from the centre account for more than 70 per cent of the total revenue of the State throughout the year since 2001. It implies that out of every rupee spent in Manipur more than 70 paisa comes from the centre in the form of dole. Manipur cannot earn even one-third of the expenses it incurs annually.

Table 7.4: Revenue Receipts of the Government of Manipur (% share)

Year	*State's Own Tax Revenue (a)*	*State's Own Non-Tax Revenue (b)*	*Share of Central Taxes (c)*	*State Own Revenue (a+b+c)*	*Grants-in-aid from centre*
1981	2.30	27.24	6.79	36.33	63.67
1985	3.53	3.04	7.45	14.02	85.98
1990	3.59	6.15	25.02	34.76	65.24
1995	4.02	8.45	30.64	43.11	56.89
2000	3.73	3.99	29.71	37.43	62.57
2001	4.70	3.99	15.65	24.34	75.66
2002	4.33	2.44	12.08	18.85	81.15
2003	4.91	4.25	14.17	23.33	76.67
2004	4.87	3.47	16.91	25.25	74.75
2005	4.67	4.01	16.47	25.15	74.85
2006	3.93	3.18	14.21	21.32	78.68
2007 (RE)	3.71	5.52	13.76	22.99	77.01
2008 (BE)	4.61	5.30	18.70	28.61	71.39

Source: Economic Review Manipur 1994-95 and Economic Survey Manipur 2006-07.
RE – Revised Estimate; BE – Budget Estimate.

Be that as it may, the issue of economic viability might have nothing to do with insurgency in the State. Besides Manipur, there are States like Meghalaya, Mizoram and Arunachal Pradesh, where insurgency is not posing any serious challenge, which also lack economic viability. The argument that we intend to put forward is that it is all the more difficult for a State to reduce the dependency while the dictates of economic growth run in contradistinction with the dictates of militancy.

Increase in the Share of Non-Development Expenditure

Looking from the point of view of the State management, increased investment in income generating economic activities can only address the issue of economic viability in the long run. This necessitates higher expenditure in development projects. This implies higher allocation for development expenditure, to use fiscal jargon.

Table 7.5: Revenue Expenditure of Manipur State Government (% share)

Year	*Development Expenditure*	*Non-Development Expenditure*
1981	73.08	26.92
1985	66.55	33.45
1990	71.52	28.48
1995	63.18	36.82
2000	58.60	41.40
2001	54.17	45.83
2002	57.97	42.03
2003	54.03	45.97
2004	57.21	42.79
2005	57.35	42.65
2006	64.12	35.88
2007 (RE)	62.69	37.31
2008 (BE)	64.23	35.77

Source: Economic Review Manipur 1994-95, Statistical Abstract Manipur 1998, Statistical Handbook Manipur 2002 and Economic Survey Manipur 2006-07.

The revenue expenditure of Manipur (Table 7.5) shows that the share of non-development expenditure is increasing over the years. This is expected in any conflict zone. Its share has increased from 26.92 per cent in 1981 to 28.48 per cent in 1990 and further to 41.40 per cent in 2000. In 2003 its share reached 45.97 per cent of the total revenue expenditure but subsequently decreased 35.88 per cent in 2006 and since then more or less remained constant till 2008.

Thus, the higher allocation for non-development expenditure necessitated due to the operation of militant groups ultimately reduces the fund allocation for development which has a strong bearing for the perpetuation of economic underdevelopment and ever rising fiscal dependence of the State on the centre. Further, the lion's share of the development expenditure is incurred in social sector leaving the real sector of the economy deprived of resources. The consequent widening gap between these two has resulted in mounting pressure of unemployment in the State which has direct bearing on the growth of militancy.

Unemployment-Insurgency Nexus

Manipur is one of the highly literate States in India. But the high literacy rate is overshadowed by the higher unemployment rate. Manipur is suffering from a chronic unemployment problem. The avenues for employment are very extremely limited due to very weak industrial base of the State. There is a mismatch between educational level and employment opportunities.

Table 7.6: Literacy Rate (percentage)

Year		Manipur	India
1981	M	64.15	56.38
	F	34.67	29.76
	P	49.66	43.57
1991	M	71.63	64.13
	F	47.60	39.29
	P	59.89	52.21
2001	M	80.30	75.30
	F	60.50	53.70
	P	70.50	64.80
2011	M	86.49	82.14
	F	73.17	65.46
	P	79.85	74.04

Source: National HDR 2001 and Statistical Abstract Manipur 2004, Census 2011.

Table 7.6 shows the literacy rate of Manipur *vis-à-vis* India in the census years. In the Census year 1981 the literacy rate was 49.66 per cent in Manipur compared to all India level of 43.57 per cent. This increased to 59.89 per cent in Manipur compared to all India level of 52.21 per cent in 1991 and further to 70.50 per cent in Manipur against all India level of 64.80 per cent in 2001. In both cases of male and female literacy rate Manipur performed well above the national average. Male literacy rate was 64.15 per cent in Manipur against the all India level of 56.38 per cent in 1981. It increased to 80.30 per cent in 2001 in Manipur against the national average of 75.30 per cent. Similarly female literacy rate in Manipur increased from 34.67 per cent in 1981 to 60.50 per cent in 2001 against the national average of 29.76 per cent and 53.70 per cent in respective years. In the Census 2011 also Manipur is far ahead than all India average. In 2011 the literacy rate of Manipur is 79.85 per cent against all India literacy rate of 74.04 per cent. In case of male literacy, it is 86.49 per cent in Manipur against 82.14 per cent of all India. Similarly female literacy rate stood at 79.85 per cent in Manipur against all India rate of 65.46 per cent (Table 7.6).

In case of adult literacy rate also Manipur is above the national average. In the year 1995-96 the adult literacy rate was 64.49 per cent against the national average of 54.32 per cent. Both male and female adult literacy rates were also above the national average. Male adult literacy rate in Manipur was 76.03 per cent against the national average of 67.25 per cent and female adult literacy rate was 52.29 per cent against the national average of 40.67 per cent.

The fact that Manipur is performing well in case of education above the national average is overshadowed by the lack of employment opportunities in the State. The State is suffering from chronic unemployment problem especially among the educated youths.

Table 7.7: Growth in Employment (Per cent per Annum)

State/Country	*1983 to 1993-94*			*1993-94 to 1999-2000*		
	Male	*Female*	*Persons*	*Male*	*Female*	*Persons*
Manipur	3.6	2.9	3.3	3.2	–0.3	2.0
India	2.2	1.7	2.1	1.9	0.9	1.6

Source: National HDR 2001.

Growth in employment is estimated as compound annual growth in the persons employed in the age group 15 years and above on the usual principal and subsidiary status. There is a decline in the growth in employment from 3.3 per cent per annum in 1983 to 1993-94 to 2 per cent per annum in 1993-94 to 1999-2000. The all India average was 2.1 per cent per annum and 1.6 per cent per annum in 1983 to 1993-94 and 1993-94 to 1999-2000 respectively (Table 7.7). The condition is worst in case of female as their growth in employment declined from 2.9 per cent per annum in 1983 to 1993-94 to –0.3 per cent per annum in 1993-94 to 1999-2000 in Manipur. The national average was 1.7 per cent per annum in 1983 to 1993-94 which declined to 0.9 per cent in 1993-94 to 1999-2000. Growth in male employment declined marginally from 3.6 per cent per annum in 1983 to 1993-94 to 3.2 per cent in 1993-94 to 1999-2000. The national average was

2.2 per cent in 1983 to 1993-94 which declined to 1.9 per cent per annum 1993-94 to 1999-2000 (Table 7.7).

The incidence of unemployment is defined as the percentage of persons unemployed in the age group 15 years and above on the usual principal and subsidiary status to the total number of persons in the labour force. The incidence of unemployment is very high in case of Manipur compared to the national average.

Table 7.8: Incidence of Unemployment (as a percentage of labour force)

State/ Country	1983			1993-94			1999-2000		
	Male	Female	Persons	Male	Female	Persons	Male	Female	Persons
Manipur	0.6	0.0	0.4	2.2	1.0	1.8	3.7	3.1	3.5
India	2.3	1.3	2.0	2.1	1.7	2.0	2.5	1.8	2.3

Source: National HDR 2001.

It increased from 0.4 per cent in 1983 to 1.8 per cent and 3.5 per cent in 1993-94 and 1999-2000 respectively. The all India figure was 2 per cent in 1983 and 1993-94 which increased to 2.3 per cent in 1999-2000 (Table 7.8).

The low growth in employment and high incidence of unemployment show the chronic unemployment problem in the State indicating a severe mismatch between the literacy rate and employment opportunities in the State. The sense of deprivation among the unemployed youths encourages them to join the bandwagon of the militancy. Many of them consider militancy as an alternative means of livelihood. Joining the struggle for social justice is often considered far better than sitting idle at home and confronting social slur. Out of sheer desperation, many start acting as the conduits before joining the militant groups as full timer. Even those who do not chart the path of militancy feel that the State is responsible for their apathetic condition and nurtures a soft corner for the so called revolutionary movements which indict the Indian State for the miseries of the people of Manipur. They become the passive supporter of militancy and at times play an active role in mobilizing social and political support for the undergrounds. Many of the educated youths become the victims of widespread corruption. In spite of being one among the equals they are often denied State privileges as they are not in a position to meet the demand for bribes. Being dejected, deprived and depressed, they turn into harsh critic of the existing political order. They fantasize about the role of the underground revolutionaries and at times fall back on them for getting justice. In the process, a symbiotic relation is established between the militant groups and the members of civil society which provides sustenance to the militancy machine.

Conclusion

The above analysis is an attempt to map the effect of insurgency on the economy of Manipur. Although it is difficult to establish any empirical association between insurgency and various economic indicators in the State, the above analysis no doubt presents a vivid picture of the overall State of the economy during these insurgency years. It has been found that the State has not progressed as much as it could have given the quantum of liberal funding from the centre and other assistance. At the same time the intensity of insurgency has shown no signs of respite. The inability of the State to create employment opportunities and its continued dependence on the centre for fund assistance has virtually trapped the State into a vicious cycle of underdevelopment and insurgency. The long years of insurgency in the State has undermined the potential of the economy which virtually transformed the State into a dependency syndrome. The State, in fact, runs with central assistance without which the State is virtually handicapped. The failure on the part of the government to properly utilize development funds partly due to interference from the insurgents and partly due to corruption further transformed the State into a fertile soil for breeding insurgents. Thus Manipur is caught in a conflict trap of underdevelopment and insurgency torn State. So any future policy intervention should aim at both reduction of the intensity of insurgency and improvement of the economic conditions of the State as a whole.

REFERENCES

Census of India. (2011). Provisional Population Tables and Annexures. Retrieved from http://www.censusindia.gov.in. (accessed August 01, 2012).

Central Statistical Organisation. (various issues). National Accounts Statistics. Retrieved from http://mospi.nic.in. (accessed August 01, 2012).

Das, Gurudas. (2002). India's North-East Soft Underbelly: Strategic Vulnerability and Security. *Strategic Analysis*. 26 (4).

Economic and Political Weekly Research Foundation, (2003). *Domestic Product of States of India 1960-61 to 2000-01*, Mumbai.

Government of India, (2002). *National Human Development Report 2001*, Planning Commission, New Delhi.

Government of India, (2006). *Basis Statistics of North-Eastern Region 2006*, Ministry of Development of North-Eastern Region, North-Eastern Council, Shillong.

Government of Manipur, (1998). *Economic Review 1994-95*, Directorate of Economics and Statistics, Imphal.

Government of Manipur, (1999). *Statistical Abstract of Manipur 1998*, Directorate of Economics and Statistics, Imphal.

Government of Manipur, (2002). *Statistical Handbook of Manipur 2002*, Directorate of Economics and Statistics, Imphal.

Government of Manipur, (2004). *Economic Survey Manipur 2002-2003*, Directorate of Economics and Statistics, Imphal.

Government of Manipur, (2005). *Statistical Abstract Manipur 2004*, Directorate of Economics and Statistics, Imphal.

Government of Manipur, (2007). *Economic Survey Manipur 2006-07*, Directorate of Economics and Statistics, Imphal.

Government of Manipur, (2008) *Economic Survey Manipur 2007-08*, Directorate of Economics and Statistics, Imphal.

Ministry of Home Affairs (MHA). (2011-12). Annual Report 2011-12. New Delhi, India: MHA.

Parratt, John. (2005). *Wounded Land: Politics and Identity in Modern Manipur*. New Delhi: Mittal Publications.

Routray, B. Prasad. (2007). Manipur: The Nexus Again. *South Asia Intelligence Review*. 6 (21).

South Asia Terrorism Portal. (various issues). Data Sheets – Manipur. Retrieved from www.satp.org. (accessed August 02, 2012).

Tarapot, Phanjoubam. (2005, 2nd Rev. ed.). *Bleeding Manipur*. New Delhi: Har-Anand Publications.

8

Post-Insurgency: Rural Development Strategies in Mizoram

Harendra Sinha

Introduction

Mizoram (earlier known as Lushai Hills/Mizo Hills District) is situated on the North-East of India located between 22°19′N and 24°19′ N Latitude and 92°16′ and 93°26′ East Longitude covering a geographical area of 21,087 sq. km. It is covered by international borders from three sides, Myanmar in the East and South (404), and Bangladesh in the West (306). Mizoram is highly mountainous and has rugged topography with high ranges trending north-south direction. Barring few patches of flat land along the valleys and the area bordering the plains of Cachar and Bangladesh, the topography of Mizoram is composed of steep hills and deep gorges (Singh 1995: 4-5). Rural development[1] in the State is not as straightforward as it is in other plain States of India. Moreover, two decades of insurgency[2] severely affected the village economy and development of Mizoram as a whole. The most peaceful State in the North-East India now, Mizoram was harshly affected in its development due to eruption of underground Mizo National Front (MNF) in midnight of 28th February, 1966 which lasted for two decades. It has disrupted not only the normal administration but also the development efforts as a whole. The grouping of villages as counter insurgency measures further aggravated the socio-economic scenario of Mizoram. This chapter focuses on the impact of Village grouping during insurgency and post-insurgency rural development initiatives in Mizoram. This chapter is organized from published and unpublished works, findings of some scholars and field work findings of the author which is required for this chapter.

This chapter is organized as follows. Section II focuses on grouping of villages and its impact on rural economy, section III discusses the post-insurgency development initiatives, section IV presents the present economic scenario of rural Mizoram and in section V, the author suggested some measures for development of rural Mizoram. All the rural development strategies in Mizoram are not covered in this paper but efforts have been made to highlight the rural development initiatives and its impact, which may serve as highlights for further research and some feedbacks for administrators and policy makers.

II

Village Grouping and its Impact on Rural Development

Occurrence of insurgency was the most significant factor which determined the course of political development in Mizoram. It has disrupted the normal administrative system and counter-insurgency measures have again affected the system in the feedback process (Ray 1982:152). Insurgency in Mizoram began in 1966 due to various factors, out of which economic backwardness and development disparity, alarming unemployment, apathy of the Assam Government, the then State Government, to solve the problem of the series of famines and crop failures, etc., led to the outbreak of insurgency. The extraordinary situation of sudden rebellion throughout the whole district resulted in the imposition of Assam Disturbed Areas Act 1955 to enable the government to make adequate provision for the suppression of the disturbances. The Act under Section 4 empowered magistrates, officers of the Assam Rifles of the rank of Havildar and upwards and police officers of the rank of sub-inspectors and above to fire upon or otherwise use force even to the causing of death, against any person who is acting in contravention of any law or in the disturbed area (*Ibid.*, 164-65). As was expected, the security forces carried out large-scale offensive operations against the MNF. Situation aggravated further when the MNF declared the 'independence' of Mizoram from

India in the early hours of March 1, 1966 followed by counter-insurgency measures when the army was asked by the Central Government to deal with the situation. As a counter-insurgency strategy, regrouping of villages was made, which saw the entire rural population of Mizoram (roughly 80% of the total population) uprooted from their homes to be relocated miles away. The Indian Army argued that the segregation and control of the population by this method was necessary for a successful counterinsurgency campaign. General Manekshaw, the Army Chief had suggested the government to hand over the district administration to the army for effective control over MNF and also suggested from military point of view that regrouping of villages should be given priority in order to control the rebels effectively (Jafa: http://www.satp.org). There were four categories of grouping of villages in Mizoram (Singh 1994:221-23).

1. **Progressive and Protected Villages (PPV):** The first stage of grouping started on 14th January, 1967 under the order of Central Government Laison Officer for Mizo Hills District. Accordingly, 50,000 people[1]. Progressive and Protected Villages (PPV): The first stage of grouping started on 14th January, 1967 under the order of Central Government Liaison Officer for Mizo Hills District. Accordingly, 50,000 people from 106 villages were resettled in 19 centres, along Silchar-Aizawl-Lunglei road which started from Vairengte villages covering a distance of 400 km. Further 63 villagers with a population of 40,000 were reorganized into 17 PPV along the sensitive borders with Burma and Manipur. Each PPV unit was placed under the charge of Civil Officer called Administrative Officer (AO), to represent the Deputy Commissioner to the people of the PPV, from Junior Assam Civil Services and a small office was provided consisting of a few clerks and porters.
2 **New Group Centre (NGC):** The second category of grouping as NGC was done immediately after completion of PPV in 1969. 182 villages into 40 group centres affecting 55,000 populations were brought under regrouping. The 40 centres were divided into 6 sectors on the three borders of the district. (i) Tripura border, (ii) East Pakistan (now Bangladesh) border, (iii) Lunglei Demagiri Road, (iv) Lunglei-Lawngtlai Road, (v) Selling-Champhai Road,

and (vi) Darlawn-Khawbung, North Vanlaiphai-Serchip Road.

3. **Voluntary Group Centre (VGC):** The third category of grouping the VGC started in 1970 and 110 villages were grouped into 26 centres voluntarily, under the supervision of the security forces with all facilities extended to them as in case of other two.
4. **Extended Loop Area (ELA):** Under this category, 63 villages were grouped in 17 grouped centres covering of total population of 40,000 along the border of Manipur in the north and Burma in the east, which was completed by the end of 1969 and the army authority handed over the area to the civil authority by the middle of 1970. There were some positive impacts of village grouping. Although public opinion in general was against grouping and all political parties wanted that the villages should be reestablished in the old sites, the grouped villages continue as such even after a popular government came in power after the election in Mizoram in 1972.There were obvious advantages of populous villages situated on the lines of communication. Administration has, therefore, preferred the people to continue to remain in the group centre, rather than moving back to scattered hamlets away from the main currents of developmental and administrative activities (Ray, 1982:177-78). This was which Tarlok Singh[3] suggested before insurgency broke out, who observed that the economic benefits could not reach the innumerable interior villages having small populations, distant and scattered hills and recommended that these small villages should be reconstituted into large units with road communication which would bring the people easy reach of technical know-how of developmental and financial and other assistance from the government (*ibid*: 169).

However, the grouping of villages had its adverse impinge on the socio-economic life of the people. The general humiliation, loss of freedom and of property, and, very often, injury and death involved in this process of so-called 'grouping of villages' were incidental to the military operations in Mizoram, as perhaps they are to internal wars anywhere. It appeared to many then, as it would appear to many more acutely now, that the policy of 'grouping' was tantamount to annihilation of reason and sensibility and certainly not the best policy to follow against our own ethnic minorities (Jafa: http:// www.satp.org). Although regrouping of villages done with the basic object of eliminating the underground elements as interior villages were under their control, yet it failed to effect physical elimination of the underground movement. It naturally disturbed the traditional village harmony based on homogeneity and attachment to village land and the masses of the village communities who were left to fight their survival with the traditional system of *jhum* cultivation became poorer (Nunthara, 1996:115). From morning to evening people had to endure the pangs of a regimented life. Their "Thalawmbawka" huts, near the sites, where *jhumias* used to stay at night, were dismantled. They had to compromise with their freedom. The man-land affinity was burnt which disturbed the geopolitical order of the area (Singh, 1994:223-25). Once self-sufficient village economy, mainly through *jhum* or shifting cultivation, had to depend on relief and temporary employment as little economic activity could be carried out under strict supervision and watchful security forces. It has its permanent repercussion even in post-insurgency Mizoram, where almost all villages and the State is not self-sufficient in food production. Till 1972, Mizoram was one of the autonomous districts of Assam. Since Assam had four other autonomous districts, the share of Mizo district was but meagre. The planning and development process, in true sense, in Mizoram started only with effect from the Seventh Five Year Plan, when the State launched her development programme for the first time under conditions of peace.

III

Post-Insurgency Rural Development Strategies in Mizoram

Post-Insurgency rural development strategies of the government have been mainly towards attaining economic self-sufficiency through agricultural development and many other segments which have potentials for local propagation and production and to protect the land from degradation, to protect burning of forests for *jhuming*, encouraging permanent cultivation in achieving sustainable development in eco-friendly manner *vis-à-vis* to find out alternative way of livelihood for the *jhumia* families. Some of the major rural development programmes initiated by the Government of Mizoram apart from the centrally sponsored schemes are:

New Land-Use Policy

The NLUP was launched with effect from the year 1990-91, put the following objectives to be achieved (GOM: 1990):

> (i) To put an end to the practice of wasteful shifting cultivation; (ii) to induce the *jhumia* families to take up an alternative permanent means of livelihood under either agricultural, industrial or animal husbandry sectors on a land allotted to them for the purpose; (iii) to take up wet rice cultivation in all the potential flat lands in order to promote self sufficiency in food production; (iv) to augment the afforestation drive; and (v) to provide marketing outlet for produces obtained from the NLUP programme.

The target group under the project includes *jhumia* families who solely depend on shifting cultivation for their livelihood and also those families other than *jhumia* families, having no permanent means of livelihood.

Under NLUP, alternative occupations were made available to beneficiaries in three sectors, viz., agriculture, animal husbandry and veterinary and industries sectors. Each beneficiary was allowed to choose one trade from out of the above sectors. Each family was given a plot of land measuring two hectares each. Each beneficiary was sanctioned an amount of Rs. 30,000 in instalments over a period of three years. The beneficiaries are required to sign an undertaking before receiving any assistance.

Performance of NLUP: Although the Presbyterian Church of Mizoram rated the success of NLUP at 40 per cent[4] of its avowed objectives, it failed during the decade period (during 1990-91 to 2001-2002) of programme implementation in its main objective to put an end to the practice of wasteful shifting cultivation and to achieve self-sufficiency in foodgrain production. Even after sanctioning more than Rs. 146 crores under the scheme, not much tangible results had been achieved as was expected (Lianzela, 2003: 146). Many people complained that expenditure of money in this manner has resulted in patronising corruption. No proper soil testing, survey, no well prepared scheme/plan was made and the scheme in many instances has appeared as mere populist measure to catch vote for the party (*ibid.*, 148). About 20 thousand *jhumia* families, out of the total of 60 thousand, were expected to settle on about 21 thousand hectares of land, thus reducing the area under shifting cultivation to the size of 56.7 thousand hectares in 1992-93. This reduction in net area affected shifting cultivation was quite insignificant. It has been noticed that most of the beneficiaries under the scheme have built their houses and brought consumer durables and they still continue with the old technique of slash-and-burn method to produce foodgrain in their fields, though relenting a little (Agarwal, 2003:106). There were allegations of too much political interference in the selection of beneficiaries, apparent lack of coordination between the government officials, the voluntary organizations and the Village Councils, the funds involved and the amount which actually reaches the beneficiaries, mismanagement of the use of fund, and also the lapses in its design (Das, 2004:46-48). According to the MS Swaminathan Research Foundation, Chennai, the poor achievement of the NLUP were due to (i) schemes chosen arbitrarily, (ii) high value schemes have been given precedence over marketability of products, (iii) poor monitoring, and (iv) purchase of raw materials, tools, animals, etc., by a Central Committee but the inputs never reached in time (Lianzela 2003:148). The programme thus, failed to produce tangible results as was expected in the decade period of implementation. With the change in government, NLUP was discontinued from the year 2001-2002 with subsequent introduction of another major project known as Mizoram Intodelhna Project during 2002-2003 under MNF Ministry headed by Zoramthanga.

Revival of New Land Use Policy: With the basic aims and objectives of NLUP intact (NLUP Manual, Detailed Action Plan: 2009), it is revived as a flagship programme for all-round development in Mizoram from 2009-2010. It is a core component of the development strategy for poverty alleviation in rural and urban areas of the New Government in pursuance to election manifesto under the Chief Minister Lal Thanhawla.

The procedure this time has been extended to 52 trades to be implemented by various development departments such as Industries Department (29 trades), Horticulture Department (9 trades), Veterinary Department (5 trades), Agriculture Department (4 trades), Soil and Water Conservation Department (3 trades), Fisheries, Sericulture and Environment & Forest Departments (1 trade each) under the umbrella of NLUP Implementing Board.

The number of households to be covered in next five years is 1.2

lakh with an average annual coverage of 24 thousands and the quantum of assistance this time is Rs. 1.10 lakh (average). A baseline survey has been made and the number of eligible families for NLUP assistance is 1.60 lakh out of total number of 2.8 lakh families surveyed (Baseline Survey: 2009). A detailed action plan of the NLUP project (both physical and financial) was made and the total funding of the project is Rs. 2907.90 crore to be approved and financed by the Government of India (Abstract, Funding Pattern NLUP: 2009) and Planning Commission of India finalized annual plan of 2010-11 for Mizoram at Rs. 1500 crore and Additional Central Assistance of Rs. 70 crore for projects of special interest (GOI: 2010).

The Prospects of NLUP (Executive Summary: 2010)

(a) **Jhum Control and Livelihood Opportunities:** It is expected that NLUP Programme as designed will progressively wean away farmers to alternative livelihood activities. Moreover the farming community with the benefit of plan intervention will be able to increase their income substantially in view of increased productivity, economics of scale and availability of markets. The project envisages creating sustained livelihood activities to an estimate 1,20,000 households over next 5 years. Projected impact on income is shown in the Table 8.1:

Table 8.1: Projected Impacts on Income from NLUP

Sl. No	*Land Use System*	*Annual Income (Rs)*	*Return to Labour Rs./day*
1.	Initial Situation	5,626	23
2.	Jhum-Agro forestry mixed G.(0.5 ha)	27,000	68
3.	Jhum –passion fruit/turmeric (0.5 ha)	19,340	56
4.	Jhum-lowland irrigated crops (0.2 ha)	13,190	37
5.	Jhum-orange/broom grass (0.5 ha)	28,280	85

Source: Executive Summary of NLUP: 2010

(b) **Food Security and Self Sufficiency:** NLUP programme focused on increased production of essential items like paddy, pulses, oil seeds, vegetables fruits, livestock, fisheries, etc., for food security and progressively attaining self-sufficiency for many items which have potentials for local propagation and production. Successful Implementation of NLUP programmes will augment the State revenue by dispensing with the need of importing such items valued Rs. 40.50 crores annually.

c) **Afforestation:** Programmes being eco-friendly and attuned to soil and water conservation would encourage of forestation on a large scale with the benefit of environmental protection and over a period time opening up scope for carbon economy.

Mizoram Intodelhna Project (MIP)

During 2002-2003, the Government of Mizoram launched a major programme known as Mizoram Intodelhna Project which may be translated as a project for self-sufficiency in Mizoram funded by Government of India for its effective implementation. On 26th March, 2002, the detailed project report was submitted to the Planning Commission of India for approval and release of funds. Subsequently, Planning Commission has agreed to release Rs. 15.00 crores as Additional Central Assistance to the Government of Mizoram. The main purpose of MIP is the enlistment of the rural poor especially the shifting cultivators. Its main concern is also in attainment of self-sufficiency in food production, food security, and better livelihood for the cultivators as a whole. The objective of the project is to develop in low land areas and terraces in the upland areas and to develop piggery for diversification on farming system. The MIP targets the shifting cultivators in selected villages only covering all the rural development blocks in the State (GOM: 2003). The MIP has failed to produce tangible results in its objective in achieving self-sufficiency in food production and in control of *jhum* cultivation in Mizoram.

Development of Agriculture

To increase production and in control of *jhum* cultivation, various works are being undertaken by the Agriculture and Minor Irrigation Department of the Government of Mizoram mainly to create potential areas for crop production and to increase yield potential per unit area by adopting new technologies, replacement of low yield local varieties of crops with high yield, construction of irrigation facilities for maximum production per unit area covering all the districts during 2002-2003.

Foodgrain Development: Under this scheme fresh flat lands as well as land improvement are done by dozer, tractor, power tiller, or manual labour, certified seeds of paddy, oil seeds, maize, pulses, etc. are introduced and made available to the farmers for sale at subsidized rate.

Fertility Management: Organic farming was introduced in the State as a result of which use of chemical fertilizers was discouraged and the bio-fertilizers are distributed to the farmers free of cost in order to encourage the farmers in its uses. A bio-fertilizers production unit was set up at Neihbawi Farm in 2001.

Plant Protection: Plant Protection chemicals are made available to the farmers at subsidized rate with Mautam or famine to occur in a very near future. The Government of Mizoram has formulated a comprehensive programme of Bamboo Flowering and Famine Combat Scheme (BAFFACOS), with an aim to reduce poverty among farmers during famine period as famine visits Mizoram at regular intervals during the flowering of *Melocanna baccifera* and *Bambusa tulda*. The occurrence of the past famines is due to the outbreak of rats, thereby causing a famine in the State. The programme of BAFFACOS adopted to achieve intensified and diversified cropping system, mechanization of agriculture, and connectivity for market linkage and Rodent Control Programmes for which total outlay of Rs. 3,300.00 lakhs was approved during 2005-2006.

Control of Jhum Cultivation: Under development component total area of 213 hectares of potential area has been developed for wet rice cultivation and 144 hectares of hillside slope, 132 hectares land improvements has also been developed for terraced cultivation. Moreover, a total length of 63 km of Potential Area Connectivity has been constructed at different locations wherever necessary. Soil conservation has been made to do away with the old practice of shifting cultivation in Mizoram which gives priority for terrace conservation under run-off scheme for creating farms. The terracing above 5 to 25 per cent slope as it is the safest for conservation of soil and moisture. The broad targets for land reclamation and development in Mizoram in the Eighth Plan (1992-97) were 8,400 hectares with a financial outlay of Rs. 400.00 lakhs.

WDPSCA and NWDP: The Watershed Development Project in Shifting Cultivation Areas (WDPSCA) and National Watershed Development Project (NWDP) are under implementation in Mizoram. The WDPSCA scheme was launched in 1994-95 in the North-Eastern States and continues during Ninth and Tenth Plan with an objective to protect and develop the hill slopes of *jhum* areas through soil and water conservation measures on watershed basis, to mitigate all effects of shifting cultivation by introducing appropriate land-use. A total of 38,08000 hectares has already completed covering 8 projects as on September 2003 during Tenth Plan. The National Watershed Development Project (NWDP) scheme was launched in 1990-91 by the Government of India with an objective of conservation, development and sustainable management of natural resources, enhancement of agricultural productivity, restoration of ecological balance and creation of sustained employment opportunities for the rural community including the landless. In Mizoram, a total of 20 and 36 numbers of project with an area of 17,666 hectares and 44,475 respectively has been treated with a total expenditure of Rs, 3,236 lakhs during Eighth and Ninth Plan period. With the implementation of WDPSCA and NWDP schemes, the potential for horticulture, species, vegetables and fodder expected to increase to a substantial extent. A Food Park is being developed at Chhingchhip village near Aizawl. Besides, the Government of Mizoram is implementing various other programmes such as Accelerated Maize Development Programme, under Technology Mission, Sustainable Development of Sugarcane Based Cropping System and Oilseed Production Programmes, Integrated Dairy Development Programme, Development of Sustainable Base for Pig Production, Rural Entrepreneurship Development Programme, etc.

However, the planned development strategies had very modest impact on the rural people and the State as a whole. In spite of the huge developmental expenditure incurred, the State is still backward with inadequate transport facilities, scarcity of drinking water, very expensive price of essential commodities, lack of marketing facilities, etc. Still today about 3 lakhs work force in Mizoram are engaged in low productivity *jhum* practice and it is estimated that an average area of 2.00 lakhs acres of forest cover are annually destroyed by slashing and burning of trees for *jhum* land in Mizoram, (http//:mizoram.gov.in), yet failed to produce sufficient food crops. The following section portrays the present economic scenario of rural Mizoram.

IV

Present Economic Scenario of Rural Mizoram

The geographical location, poor condition of road networks,

remoteness of the market, poor infrastructural facilities, no broad gauge rail road and waterways to transport goods to outside world, slow growth of alternative employment avenues, lack of industries, deficiency in natural resource management, inappropriate programmes and policies initiated and poor implementation of development programmes are some of the evils towards development in Mizoram. Despite its abundant natural forest resources, favourable climate and a fertile land, the State remains backward. It has been stated that, "during the sixty years of planned development, India has made impressive progress in developing agriculture and industry, science and technology, health and education and the infrastructure for a wide range of services". But it is sad to note that this statement has no relevance so far as Mizoram is concerned. Mizoram is still virtually where it was at the time of attainment of Union Territory. No doubt, the rest of the country has made impressive progress, but Mizoram is still far behind even in building up the requisite infrastructure (Hluna, 2010: 4). The land is fertile and favoured by monsoon. Its forests are abundant green (Das, 2004:33). But in spite of its natural resources, Mizoram remained socio-economically backward. The region is at an infantile stage in the field of industry. Whatever industry, it may be called; at present is only small-scale industry, fed by indigenous agricultural produce covering only a small market zones (Pachuau, 1994: 73). The situation in respect of allied sector is in no way better. The potential in animal husbandry, a common subsidiary activity, is grossly underutilized and meat, eggs, fish come from outside the State (*ibid*: 34). Agriculture still remains underdeveloped and the primitive method of *jhum* predominates, while only 5 per cent of the total area is under cultivation. Paddy continues to remain the chief food crop and the staple food of the Mizo. It occupies almost 50 per cent of the total cropped area and more than 88 per cent of the total area under foodgrains. In spite of the fact that the rice being the most important crop occupying the largest share in area and production, is still not self sufficient in rice production where about 75 per cent[5] of rice alone comes from outside the State.

The share of agriculture alone in the net domestic product (NSDP) is merely 30 per cent at current price during 1998-99. As per the Agriculture Census (1995-96), there were 65,919 operational holdings with a total operated area of 85,000 hectares. Out of the total number of holdings 42.04 per cent is marginal 39.0 per cent small, 17.83 per cent semi-medium 1.11 per cent medium and 0.01 per cent large holdings. Mizoram an agrarian economy still imports a large quantity of foodstuff, fruits, vegetables, fish and livestock like pig, cattle, goat, poultry essentially required for the overall food supply of the people. Large quantities of such items are presently imported from other States particularly neighbouring States of Assam, Tripura and Manipur. Some quantity is also imported through informal border trade from Myanmar. As per the Trade and Commerce Department, Mizoram, the value of Import of Agriculture, Horticulture and Livestock's items, etc., during 2008-2009 was Rs. 4,04,91,1730.00, i.e., Rs. 40.50 crores (http//:mizoram.gov.in).

According to the 1990 Survey Report of the Rural Development Department, 52 per cent of people today in the State were living below poverty line. It is a paradoxical situation that, more than ever the threats of poverty loom large in the hearts of the working class despite the fast growth in developmental expenditures since 1972 (Thangchungnunga, 2003: 143). With the existing structure of Mizoram economy, precarious State of agriculture slow growth of alternative employment avenues in rural areas, lack of industries, poor communication facilities, there is low possibility of absorbing the fast expanding labour force. The cultivable land and other traditional rural enterprises have a limited scope for supporting the fast emerging population. This has forced more people to seek their fortunes in the urban areas, which have its own repercussions on the economy (Thansanga, 2003: 215-16). Table 8.2 show the total area of land utilisation in Mizoram.

Table 8.2: Land Utilisation in Mizoram during 2006-07

Sl. No.	*Area under*	*In '000 Hectares*	*Percentage of total*
1.	Forest	1593.700	75.57
2.	Not available for cultivation	134.040	6.37
3.	Other cultivated land	79.230	3.76
4.	Fallow land	207.543	9.84
5.	Net sown area	94.187	4.46
6.	Total	2108.700	100.00

Source: Directorate of Agriculture and Minor Irrigation, Government of Mizoram, Aizawl.

The total geographical area of the State is 21,08,700 hectares in which 9.84 per cent is under fallow land and only 4.46 per cent is under net sown area. Although the administration was geographically closer to the rural people due to grouping of villages, it failed to create sustained impact on the lives of the rural people. This can be easily blamed for insurgency. But rural bureaucracy is hitherto entrenched to the rural people. The following Table 8.3 shows the responses on whether Block level bureaucrats are visiting the villages regularly. 75 per cent of the respondents says 'no', 4.17 per cent responded 'yes' and 20.83 per cent falls in the category 'do not know' (Sinha, 2006: 230).[6]

Table 8.3: Showing Responses on whether the Block Level Bureaucrats visits the Village People regularly

Number of Respondents—240

Block Functionaries	*Category*	*No. Responding*	*Percentage*
BDO	Responding 'Yes'	10	4.17
EOs	Responding 'No'	180	75.00
VLWs	Do not know	50	20.83
Total		240	100.00

Source: Questionnaire and Interview Schedule.

The message from Table 8.3 replicates the non-availability of the rural bureaucracy in the villages. Many remote villages were not visited by these functionaries although it was found that the villagers heard about these functionaries. There is lack of motivation and belongingness of all rural bureaucrats at the State and Block level. This is because almost all bureaucrats are on deputation from development departments who likes to go back to their parent departments.[7]

The post-insurgency major rural development programmes of both Central and State Sponsored Schemes initiated such as Integrated Rural Development Programme, Swarnajayanti Gram Swarojgar Yojana, Border Area Development Programme and New Land-Use Policy, Mizoram Intodelhna Project, etc., failed to produce substantial outcomes and the State is far away from attaining economic self-sufficiency. Inequality in economic status in Mizo society has been very much evident. While launching New Land-Use Policy by dissolving the Mizoram Intodelhna Project, the Chief Minister of Mizoram Lal Thanhawla sounded a warning that if the present trend of widening gap between the rich and the poor is not reversed, there may be discontent amongst the poorer sections leading to social tensions, disharmony and violence to the extent of causing bloodshed in Mizoram.[8]

V

Suggestions

Lack of good communication facility is the major problem faced by the rural population of the State. Lack of road transport facilities, the farmers of the remote villages cannot market their products even to the Block headquarters. Regulated markets and cold storage facilities in the State are inadequate. Although the agro climatic condition of the region is conducive, dairy, poultry, fishery, etc. has not been developed as a means of livelihood. Veterinary services in the rural areas are inadequate. The efforts of the government in developing small-scale industries are not quite satisfactory and also this sector has a direct link with road transport, power, etc. Middleman's share has been one of the main reasons behind increasing food prices at the retail level. How to eliminate the middleman and thus cut costs and to reach the actual profit to the rural development beneficiaries were generally not taken into consideration while implementing rural development schemes. Therefore, effective implementation of development initiatives is vital. Designing and implementing people-centric schemes by harnessing the available natural resources of the State through participatory development strategy and grassroots planning is desirable. More infrastructure development ensuring better link roads, market connectivity, cold storage facilities, irrigation are required. Initiatives for more farmers' cooperative societies should be encouraged, which will help in eliminating the middleman. Making *jhum* cultivation scientific, understanding of the technical factors involved while implementing development programmes, timely supply of appropriate inputs, flow of assistance, marketing of outputs, strict and continuous monitoring are essential. Bureaucracy needs be very active and work on the field in reaching the remote villages and see that they do not look at the problems from the office. Demonstrations at regular intervals to impart technical know-how and skill to the farmers through training, tours, audio-visuals, seminars to aware and in involving the farmers to join actively in the programme is needed for which trained, skilled, specialised, efficient, impartial and more importantly motivated bureaucracy is required.

Providing gainful and productive employment opportunities is essential for which capacity building focusing on development of skills and training is extremely important. Local means and production suited to the genius of the people needs to be identified and explored. Industrial development programmes announced by the government from time to time lack efficiency in management and there is lack of proper coordination among various agencies in implementation of the programme (Laskar, 2010: 132-33). Therefore, small-scale industries based on local resources like agricultural products, bamboo and other forest products needs to be identified with special packages of incentive for educated youth to take those enterprises. Involvement of Non-Government Organisations (NGO) in providing loans in such sectors as floriculture, small-scale industries like development of weaving and knitting, horticulture, etc., and help in the marketing of those products, will ensure employment to a sizeable number of poor people.

Development initiatives for the rural people and autonomy should go together. The local self-government institutions—the Village Councils are not as viable as Panchayati Raj Institutions (PRI) due to lack of financial autonomy. Therefore, empowering the rural people by strengthening of local self government system through PRIs as per the Constitution (73rd Amendment) Act at the Village and District level with certain modifications in ensuring the existing traditional values will guarantee financial autonomy, reservation of women and the decision making power (Sinha, 2010: 47).

Concluding Remarks

The massive development initiatives undertaken by both the Central Government and the State Government, so far benefited only a small elite section of the society in the State and the rural marginal section of the society is largely deprived. Most of the rural development plans and strategies were not made in accordance with the needs of the newly uprooted groups of villages. The strategies of rural development by the Central Government such as IRDP, DWCRA, NREP, RLEGP were made for the whole of the nation and there was no systematic and sustained development initiatives were made especially for the displaced people and for their newly settled villages in the State. The result is that almost all villages and the State was not self-sufficient just after grouping and the trend continues till today. Insurgency being a deviation of the past, the State has great future to look ahead for which a corruption free responsive administration is necessary. The development projects and schemes should be prepared to fit in a methodical and inclusive plan which promises to benefit the poor people. At the same time people's participation for successful implementation of schemes and their hard work without waiting for help from the government every time, is also essential for development of rural Mizoram towards achieving economic self-sufficiency.

NOTES

1. Rural Development Sector Policy Paper, World Bank, Washington D.C, 1975, defines rural development as "the strategy designed to improve the economic and social life of the specific group of people—the rural poor. It involves extending the benefits of development to the poorest among those who seek a livelihood in rural areas—the group includes small scale farmers, tenants and landless.
2. Online Encyclopedia Wikipedia defined 'insurgency' as the state or condition of being in revolt or insurrection; an organized movement aimed at the overthrough of a constituted government through use of subversion and armed conflict.
3. Tarlok Singh, a Member of the Planning Commission, led a Study Team to the Mizo Hills in January 1966.
4. Presbyterian Church of Mizoram had rated success at 40 per cent of the objective of NLUP (Minute, Matter of Dissolution of MIP and Launching of NLUP, 5 February 2010).
5. In reply to the Assembly House Members, Agriculture Minister of Mizoram, H. Liansailova said that the production of rice could substantiate only about 25 per cent of the State's requirements (huge decrease in production due to famine (*mautam*) during 2007-2008 (Virthli: March 2010: 1).
6. Total of 120 beneficiaries + 120 non-beneficiaries of rural development, covering all the four Blocks of Lunglei District of Mizoram through questionnaire and interview schedule was made during 2003-05.
7. Based on Personal Interview with Dr. R. Lalmalsawmi (DD,Vety.), Directorate of Rural Development, Aizawl on 23.10.2005
8. Speech of Mizoram Chief Minister Lal Thanhawla at the meeting on Matter of Dissolution of MIP and Launching of NLUP in Mizoram (GOM 2010:2).

REFERENCES

Das, K. (2004) *Rural in Mizoram*. New Delhi: Mittal Publications

Fambawl, L. (1998) Mizoram—A Rural Development Department In Kum 50 chhung a Hnathawh Tlangpui. *NLUP Sekibuhchhuak'—A Bulletin of Commissionerate of Rural Development*, Vol. 31 (4).

(GOM) Government of Mizoram (1989). *Mizoram District Gazetteers*. Aizwal: Department of Arts & Culture.

—— (1990). *Project Report on NLUP*. Aizwal: Commissionerate of Rural Development, Aizawl.

—— (1998). *Directoı ate of Agriculture and Minor Irrigation*. Aizawl: GOM.

—— (2003). *Statistical Abstract, Directorate of Agriculture and Minor Irrigation*. Aizawl: GOM.

—— (2010). *Executive Summary, NLUP*. Aizawl: GOM.

—— (2010). *Minute of Matter of Dissolution of MIP*. Aizwal: Department of Agriculture.

GoI (Government of India), (2010). *Press Information Bureau, Mizoram Annual Plan Finalised, 22 March*. Aizwal: GOM.

Hluna, Dr. J.V. (2010). "Socio-economic Development in Mizoram-Problems and-Prospects" – A seminar paper presented in UGC Sponsored State Level Seminar-organized by Govt. J. Buana College, Lunglei, Mizoram, 22-23rd September, 2010.

Official website of Mizoram Government. http//:mizoram.gov.in [accessed on August 2, 2011].

Jafa, V. Singh (2011). *Counterinsurgency Warfare—The Use and Abuse of Military Force,* Retrieved from *http://www.satp.org* [Accessed on August 2, 2011].

Laskar, B.I. (2010). Industrialisation of Mizoram: Need, Opportunities, Growth and Constraints of Small Scale Industries" in Laskar, B.I. (ed), *Mizoram: The Emerging Issues of Development*. Guwahati: DVS Publishers.

Lianzela (1994). *Economic Development of Mizoram*. Guwahati: Spectrum Publications.

——(1995). *Four Decades of Planning in Mizoram*. Guwahati: Spectrum Publications

McCall, A.G. (1949). *Lushai Chrysalis*. London: Luzac & Co.

Nunthara, C. (1996). *Mizoram Society and Polity*. New Delhi: Indus Publishing Company

Pachuau, R. (1994). *Geography of Mizoram*. Aizwal: R.T. Enterprise.

Ray, A. (1982). *Mizoram Dynamics of Change*. Calcutta: Pearl Publishers

Singh, S.N. (1994). *Mizoram*. New Delhi: Mittal Publications.

—— (1995). "Mizoram: A Study in Geopolitical Perspective", in Prasad,

R.N. and Agarwal A. K. (ed) *Landmarks: A Study of Public Administration in Mizoram.* Aizawl Branch, Mizoram: IIPA.

Sinha, H. (2006). *"The Role of Block Development Officer in the Rural Development of Mizoram with Special Reference to Lunglei District Blocks (Ph.D thesis).* Gauhati: Guahati University.

——, (2010): Block Administration without Panchayati Raj Institutions—An Assessment in Lunglei District of Mizoram. *Kurukshetra,* Vol. 53 (5).

Thangchungnunga (2003). Agrarian Change and Social Transformation among the Mizos, in Prasad, R.N. and A.K. Agarwal (ed.) *Modernisation of The Mizo Society.* New Delhi: Mittal Publications.

Thansanga, H. (2003). Poverty and Unemployment in Mizoram, in Prasad, R.N. and A.K. Agarwal (ed) *Modernisation of The Mizo Society,* Mittal Publications, New Delhi.

9

Issues and Challenges of Primary Education in Tripura: Drop-Out is still a Major Challenge

Anindita Sengupta

Introduction

Tripura is a small State located on the extreme north-east corner of India. After India's independence, the princely state of Tripura was merged with the Union of India in 1949. Tripura became a Union Territory on 1 July, 1963 and attained the status of a full-fledged State on 21 January, 1972. The Department of Education in this State was established to enable smooth functioning of the education system at all levels. The department was further trifurcated to three separate sections namely: Education (School), Education (Higher) and Social Welfare & Social Education Department. Being a signatory of the World Conference of "Education For All" in Jomtien, Thailand (1990) and later on in Dakar, Senegal (2000), India is committed to achieve Universalisation of Elementary Education (UEE) in all States, as a part of the goal of Education For All (EFA) by the year 2015. To achieve this goal of UEE, over the years, a number of schemes and programmes have been launched in pursuance of the emphasis embodied in the National Policy of Education (1986, 1992) and the Programme of Action (1992). These included the State Specific Basic Education Projects like the Andhra Pradesh Primary Education Project (APPEP); Bihar Education Project (BEP), Uttar Pradesh Basic Education Project, Lok Jumbish (LJP) in Rajasthan; National

Programme of Nutritional Support to Primary Education (MDM); District Primary Education Programme (DPEP), Sarva Shiksha Abhiyan (SSA), Education Guarantee Scheme, etc. Most of these programmes were meant only to cover one particular State. Among all these programmes, District Primary Education Programme (DPEP) and Sarva Shiksha Abhiyan (SSA) cover more than one State. But ironically, before 2001-2002, no special efforts in the form of programmes were made to achieve goal of UEE in the North-Eastern States. Sarva Siksha Abhiyan has been the first serious effort towards achieving the goal of UEE in these States (Zaidi, 2005). The programme had been launched in the States of North-East in 2001-02. The SSA programme had been launched in the four districts of the State of Tripura in 2001-02. Since then, there has been significant increase in the number of primary level schools in the districts of the State. However, during the period from 2005-06 to 2009-10, in all the districts, numbers of government schools have been found to be much higher than the number of private schools for primary level, primary and upper primary level and primary, upper primary and secondary/higher secondary level. Among the four districts, West Tripura has been found to have maximum number of both government and private schools in all the levels throughout the whole period. In terms of the number of government schools of all levels, South Tripura has been found to occupy the second position and in terms of private schools of all levels, North Tripura has been found to occupy the second position during the same period. Table 9.1 also shows that while the number of government schools has increased in all the districts over the years, number of private schools has not increased in all the districts and for all levels.

If we examine the progress of enrolment of students in the primary level educational institutions in the districts of Tripura, we find that in all four districts share of enrolment has been much higher in the rural schools than that in the urban schools, although in case of primary schools with upper primary and secondary/higher secondary classes enrolment in the urban areas have been found to be slightly higher than in case of the other two types of primary schools (see Table 9.2). If the nature of enrolment of students in the primary level institutions of Tripura is examined during the period of the study, it is found that in all the districts the Gross Enrolment and Net Enrolment Ratio have been very high throughout the whole period of study (see Table 9.3). Therefore, it can be concluded that in terms of enrolment of students

Table 9.1: Primary Schools, Primary with Upper Primary Schools and Primary with Upper Primary and Secondary/Higher Secondary Schools in the Districts of Tripura

Year	Districts/ Schools	Primary only		Primary with Upper Primary		Primary with Upper Primary and Secondary/Higher Secondary	
		Government	Private	Government	Private	Government	Private
2005-06	Dhalai	363	1	154	2	47	2
	North	238	12	273	6	96	10
	South	601	1	225	1	134	9
	West	635	21	331	13	266	36
2006-07	Dhalai	389	2	154	1	47	2
	North	251	13	271	6	97	9
	South	635	4	223	4	133	10
	West	682	22	336	12	268	36
2007-08	Dhalai	432	6	162	3	48	2
	North	266	13	264	6	105	9
	South	672	5	222	5	149	9
	West	730	30	347	16	281	43
2008-09	Dhalai	403	6	191	3	48	2
	North	265	10	265	8	105	10
	South	662	4	233	5	149	10
	West	724	30	351	17	285	43
2009-2010	Dhalai	501	6	211	3	51	2
	North	330	14	256	6	115	8
	South	736	6	254	4	161	11
	West	762	35	391	11	309	45

Source: Mehta (2006-2010)

in the primary level institutions, all the districts of Tripura have done the perfect job throughout the period of the study.

According to the Census of India report in 2001, 17.4 per cent and 31.1 per cent of total population of Tripura are Scheduled Castes and Scheduled Tribes, respectively. On the other hand, according to DISE data, while the percentage of SC enrolment out of total enrolment in the primary educational institutions of Tripura have been 18.5, 18.9, 18.3, 19 and 18.5, the percentage of ST enrolment out of total enrolment in the primary educational institutions of the State have been 41.5, 41.1, 42.1, 43 and 43.5 in 2005-06, 2006-07, 2007-08, 2008-09 and 2009-10, respectively (see Table 9.4). Therefore, it is evident that the enrolments of Scheduled Caste and Scheduled Tribe

Table 9.2: Share of Enrolment in Total Enrolment in Primary Schools, Primary With Upper Primary Schools and Primary with Upper Primary and Secondary/Higher Secondary Schools in the Rural and Urban Areas of the Districts of Tripura

Year	*Districts/ Schools*	*Primary Only*		*Primary with Upper Primary*		*Primary with Upper Primary and Secondary/Higher Secondary*	
		%Share of Rural Enrolment in Total Enrolment	*%Share of Urban Enrolment in Total Enrolment*	*%Share of Rural Enrolment in Total Enrolment*	*%Share of Urban Enrolment in Total Enrolment*	*%Share of Rural Enrolment in Total Enrolment*	*%Share of Urban Enrolment in Total Enrolment*
2005-06	Dhalai	98.80	1.20	97.93	2.07	88.75	11.25
	North	93.40	6.60	97.68	2.32	80.02	19.98
	South	94.84	5.16	96.97	3.03	92.40	7.60
	West	91.98	8.02	90.49	9.51	66.58	33.42
2006-07	Dhalai	97.73	2.27	97.93	2.07	98.80	1.20
	North	94.17	5.83	98.23	1.77	93.42	6.58
	South	93.50	6.50	97.57	2.43	94.84	5.16
	West	92.80	7.20	89.91	10.09	91.98	8.02
2007-08	Dhalai	96.98	3.02	97.67	2.33	90.32	9.68
	North	94.32	5.68	96.94	3.06	80.74	19.26
	South	95.48	4.52	97.47	2.53	90.87	9.13
	West	91.37	8.63	90.56	9.44	68.54	31.46
2008-09	Dhalai	97.62	2.38	97.77	2.23	88.70	11.30
	North	93.51	6.49	97.04	2.96	81.81	18.19
	South	96.03	3.97	95.90	4.10	91.25	8.75
	West	91.18	8.82	90.51	9.49	69.16	30.84
2009-10	Dhalai	98.77	1.23	96.57	3.43	88.93	11.07
	North	94.22	5.78	97.51	2.49	82.82	17.18
	South	96.55	3.45	97.33	2.67	88.97	11.03
	West	88.06	11.94	90.01	9.99	67.71	32.29

Source: Author's Calculation on the basis of data from Mehta (2006-2010).

students in the primary educational institutions have been quite high in Tripura in the past five years.

Although Tripura has done a perfect job in terms of enrolment of students in the primary level educational institutions under the SSA programmme, but to find out whether this effort has improved the situation of primary education in the State, we have to examine the performance of the students in the primary level. While promotion rate can be treated as the proxy variable of the performance, repetition rate and the drop-out rate can be considered as the serious obstacles

Table 9.3: Gross Enrolment Ratio and Net Enrolment Ratio in the Districts of Tripura

Year	*Items/ Districts*	*Dhalai*	*North*	*South*	*West*
2005-06	GER	140.9	124.6	151.1	125.9
	NER	100.0	100.0	100.0	100.0
2006-07	GER	139.9	132.4	158.4	127.8
	NER	100.0	100.0	100.0	100.0
2007-08	GER	161.0	140.2	145.1	122.3
	NER	100.0	100.0	100.0	100.0
2008-09	GER	175.0	131.9	150.7	123.6
	NER	100.0	100.0	100.0	100.0
2009-10	GER	175.0	133.3	158.3	121.8
	NER	100.0	100.0	100.0	100.0

Source: Mehta (2006-2010)

Table 9.4: Enrolment of Scheduled Caste and Scheduled Tribe Students in the Primary Level Educational Institutions in Tripura

Year	*% of SC Population*	*% of SC Enrolment in Total Enrolment*	*% of ST Population*	*% of ST Enrolment in Total Enrolment*
2005-06	17.4	18.5	31.1	41.5
2006-07	—	18.9	—	41.1
2007-08	—	18.3	—	42.1
2008-09	—	19.0	—	43.0
2009-10	—	18.5	—	43.5

Source: (i) Census of India , 2001 and (ii) Mehta (2006-2010).

to the educational progress of the students. In Dhalai district while the promotion rate has increased from 79.2 per cent to 81.9 per cent during the period from 2005-06 to 2009-10, drop-out rate has also increased 10.4 per cent to 16.3 per cent during the same period, although repetition rate has declined from10.5 per cent to 1.9 per cent during the same period. In North district, while promotion rate has increased from 81 per cent to 90.6 per cent during the period from 2005-06 to 2009-10, drop-out rate has decreased from 9.8 per cent to 6.5 per cent and repetition rate has also declined from 9.3 per cent to 2.9 per cent during the same period of time. In South district while the promotion rate has increased from 80.9 per cent to 89.9 per cent, drop-out rate has increased from 0.2 per cent to 4.9 per cent, although repetition rate has declined from18.9 per cent to 5.2 per cent during the same period. In West district, while promotion rate has increased from 81.3 per cent to 95 per cent during the period from 2005-06 to 2009-10, drop-out rate has also increased from 1.9 per cent to 2.2 per cent, although repetition rate has declined from 16.7 per cent to 2.8 per cent during the same period of time. (see Table 9.5)

Table 9.5: Repetition Rate, Drop-out Rate and Promotion Rate in the Primary Level Educational Institutions in the Districts of Tripura

Year	*Items/Districts*	*Dhalai*	*North*	*South*	*West*
2005-06	Repetition Rate	10.5	9.3	18.9	16.7
	Drop-Out Rate	10.4	9.8	0.2	1.9
	Promotion Rate	79.2	81.0	80.9	81.3
2006-07	Repetition Rate	7.2	5.5	11.0	11.7
	Drop-Out Rate	12.4	11.7	6.6	4.6
	Promotion Rate	80.3	82.8	82.4	83.7
2007-08	Repetition Rate	6.1	3.8	6.1	3.2
	Drop-Out Rate	4.9	9.2	16.6	8.7
	Promotion Rate	89.0	86.9	77.3	88.1
2008-09	Repetition Rate	8.0	3.6	5.6	2.8
	Drop-Out Rate	2.0	14.1	4.1	2.2
	Promotion Rate	92.0	82.3	90.3	95.0
2009-10	Repetition Rate	1.9	2.9	5.2	0.9
	Drop-Out Rate	16.3	6.5	4.9	5.4
	Promotion Rate	81.9	90.6	89.9	93.7

Source: Mehta (2006-2010).

Therefore, it can clearly be said that since the inception of SSA programme, Tripura has achieved remarkable success in terms of enrolment of students in the primary level educational institutions, especially in rural areas of the districts. But the goal of the UEE cannot be achieved through successful enrolment only. It requires successful promotion of the students to the higher level of education. In this respect, Tripura has still not been able to achieve significant success. Although the rate of promotion in all the districts has increased and the rate of repetition has also decreased in the past five years, but drop-out rate has increased in almost all the districts during the same period of time. This implies that the problem of drop-out is the major challenge of primary level of education in the districts of Tripura. Drop-out represents a staggering loss and a significant obstacle in achieving the goal of UEE. Although it is found at both primary and secondary levels, it is more crucial at the primary level.

An early school drop-out soon lapses to illiteracy. Re-entry to the formal school system is closed off in most cases. A secondary school drop-out has at least acquired literacy and some other educational skills and knowledge. He or she may more easily resume schooling, or take up training outside the system. The primary school drop-out will remain locked into the closed world of the illiterate and thus have further restriction placed upon a background of poverty and ignorance. There may be many reasons for dropping out of students in the primary level of education in Tripura. Firstly, discrimination against ST and SC children is quite evident in the education system of India. It has already been mentioned that the enrolments of Scheduled Caste and Scheduled Tribe students in the primary educational institutions have been quite high in Tripura in the past five years. Scheduled Caste and Scheduled Tribe people are the most vulnerable and poorest section of the society. Often it is found that the survival of these poorest families depends on the children's earnings. Even if the children of these poor backward class people are enrolled in the primary schools, the pressure of running the family compels the parents to withdraw their children from the schools after few years and send them to work. Therefore, high percentage of enrolment of Scheduled Caste and Scheduled Tribe students in the primary level may result in a situation where many of these students drop-out after a few years. Secondly, children from poor families, children from the families of backward class people and the girl children are the indispensable part of those students who are enrolled in the government run primary schools. These children are psychologically weak and vulnerable and, therefore, each of them require special attention and care from the teachers in order to adjust the new environment and develop a linking for the schools. If classrooms are not insufficient in the primary schools compared to the number of students, i.e., if the classrooms are overcrowded, it becomes nearly impossible for the teachers to take care of each and every student separately. In such a situation, lack of personal care discourages the children to remain connected with the schools and results in early drop-outs. Thirdly, girls can be assumed to be a large proportion to drop-outs. In almost every State of India, education is still not considered necessary for girls and they are kept at home in order to do the household chores and are prepared to get married soon. Therefore, even if the girls are enrolled in the primary schools, after a few years they drop out from there. There are some other reasons for dropping out of girls from the primary schools, e.g., non-availability of enough number of separate schools for girls and non-availability of women teachers in the schools. Fourthly, proper use of teaching aids helps to retain more concepts permanently. Through teaching aids, the teacher clarifies the subject matter more easily. Teaching aids make the classroom live and active and thereby create the environment of interest for the students. The blackboard is probably the most convenient, effective and widely used basic visual teaching aid. Neatness and orderly arrangement of blackboard material is eye catching and contribute to increase in memory. The methodical use of blackboard keeps up tempo of the class. The students watch teacher's movements. They remember and imitate what he/she does. Therefore, non-availability of blackboard in a school may be supposed to be a significant cause of dropping out of children in the primary level. Finally, it is observed that among drop-outs there are a majority of repeaters, which reflects the poor achievement of children particularly in earlier classes. Some psychologists as Jimerson, Carlson, Rotert, Egeland, and Sourie (1997) consider that early grade repetition has a negative effect on socio-emotional adjustment. Hence, one can expect grade repetition to be discouraging and cause drop-out. Grade repetition may be discouraging for a least two reasons. First, it extends the time needed to achieve a given final grade and get the benefits from education. So grade repetition may increase the cost of schooling: for a given last grade attended, the opportunity costs increase by one year when a child has to repeat once, and the job market benefits of schooling are postponed by one year. Second, grade repetition may be a negative signal about a child's ability. If the parents observe their children's ability noisily, then grade repetition diminishes parents' belief in their children's ability. Therefore, it can safely be assumed that the higher the repetition rate the higher will be the drop-out rate in the primary level of education.

Let us see what the literature on drop-out problem tells us. There is indeed a plethora of theoretical literature pertaining to the problem of drop-out, especially in the school level. Rumberger (1983) conducted a study on the extent of the high school drop-out problem and investigated both the reasons stated behind the decision of dropping out by the students and some of the underlying factors influencing their decision. The study showed that students leave school for a variety of reasons, with women more likely to leave because of pregnancy or marriage and men more likely to leave to go to work. The study also indicated that the family background strongly influenced the propensity to drop-out of school and accounted for virtually all the racial differences in drop-out rates. The study further pointed out that various other factors, including ability and aspiration, also influenced the decision of dropping out. In a study on the early school drop-out, B.R. Cairns, D.B. Cairns and J.H. Neckerman (1989) tried to examine the behavioural, cognitive, and demographic factors associated with early school drop-out. The results showed that the males and females most vulnerable to early school drop-out were characterized by high levels of aggressiveness and low levels of academic performance. Socio-economic status, race, and early parenthood were also found to be associated with school drop-out. R.W. Rumberger and A. K. Larson (1998) did a study on the incidence of student mobility between the eighth and twelfth grades and its effect on high school completion using the National Educational Longitudinal Survey third follow-up data. Three models were tested on two groups of students. The models were developed from a conceptual framework based on theories of dropping out, post-secondary institutional departure, and student transfer adjustment. The results suggested that low grades, misbehaviour, and high absenteeism were the most significant explanatory factors of both the problem of dropping out and school mobility. The results further indicated that, controlling for other predictors, students who made even one non-promotional school change between the eighth and twelfth grades were twice as likely not to complete high school as students who did not change schools. Together, the findings suggested that student mobility was both a symptom of disengagement and an important risk factor for high school drop-out. S. Jimerson, B. Egeland and L.A. Sroufe (2000) examined the relation of a variety of demographic, school, individual and family characteristics to high school drop-out. The results of this study demonstrated the association of the early home environment, the quality of early care giving, socio-economic status, intelligent quotient, behaviour problems, academic achievement, peer relation and parent involvement with dropping out of high school at age 19. These results pointed out that the psycho-social variables prior to school entry, e.g., the early home environment, the quality of early care giving, socio-economic status were equally important in explaining the drop-out problem as the factors like intelligence quotient and school achievement test scores. Ash Louise (2007) conducted a study for assessing the number of out of school children in India in the age group 6-13. The survey was conducted in all the States and Union Territories covering rural and urban areas of 588 districts. The findings of the survey indicated that the country had about 19.4 Crores children in the age group 6-13 (i.e., 6 to below 14 years) in 2007, of whom 6.94 per cent children were out of school. The study also pointed out that amongst the out of school children, 68.3 per cent children never attended school and 31.7 per cent were drop-outs. Arindam Banik and Debasis Neogi (2011) conducted a study on the problem of drop-out of children from primary schools in the tribal dominated areas of North-East India. The study found interesting results based on logit model. Factors such as earning compulsion at the household level, early marriage, and teacher not belonging to the same tribal community appeared to be the dominating variable. The study also pointed out that the girl children dominated in drop-out rates. From the literature survey it can be concluded that low levels of academic performance, socio-economic status, race, misbehaviour and high absenteeism, school, individual and family characteristics had been the most significant explanatory factors of drop-out. The studies also pointed out that the girl children dominated in drop-out rates. However, in case of Tripura, there is as such no study on the drop-out of the school students, especially at the district-level, i.e., taken all the districts together. Under this backdrop, an intensive study on district-level drop-out problem in the primary level educational institutions of Tripura is quintessential. So this chapter tries to find out the proximate explanatory factors behind the drop-out of students in the primary level educational institutions of Tripura. I have tried to estimate the drop-out function of primary level students in terms of a panel data regression model for the last five years, i.e., for the period from 2005-06 to 2009-10. This chapter is structured as follows. Section II explains the methodology and data; section III presents the results of the econometric analysis and finally section IV gives the concluding observations.

II

Methodology and Data

In this analysis, drop-out is construed as that situation where children who enrol themselves in schools but withdraw prematurely before they complete the stage.

$$\text{Drop-out Rate } (drp_t^g) = \frac{DRP_t^g}{ENR_t^g} \times 100$$

Where, (DRP_t^g) = Number of students dropping out from grade '*g*' in year '*t*'.

ENR_g^t = Total number of students enrolled in grade '*g*' in year '*t*'.

It is assumed that drop-out of students depends on a variety of factors and the drop-out function of students of primary grade (I-V) can be written as:

$$(drp^g{}_t) = f(SC_t^i, ST_t^i, SCR_t^i, GPT_t^i, WTBB_t^i, rept_{t-1}^i) \quad ...(1)$$

Where, year '*t*' = 2005-6, 2006-07, 2007-08, 2008-09, 2009-2010,
District '*i*' = 1, 2, 3, 4,

$drp^i{}_t$ = Drop-out rate in the primary classes in the *t*-th year in the *i*-th district

$SC^i{}_t$ = Percentage of SC enrolment in primary classes in the *t*-th year in the *i*-th district
(Enrolment of SC students in primary classes in the *i*-th district/Total enrolment in primary classes in the *i*-th district) × 100

ST_t^i = Percentage of ST enrolment in primary classes in the *t*-th year in the *i*-th district
= (Enrolment of ST students in primary classes in the *t*-th year in the *i*-th district/Total enrolment in primary classes in the *t*-th year in the *i*-th district)× 100

SCR_t^i = Student-classroom ratio in primary classes in the *t*-th year in the *i*-th district
= Total enrolment in primary educational institutions in the *t*-th year in the *i*-th district/Total classrooms in primary educational institutions in the *t*-th year in the *i*-th district

GPI_t^i = Gender Parity Index in primary classes in the *t*-th year in the *i*-th district
= Girls' enrolment in the primary classes in the *t*-th year in the *i*-th district/Boys' enrolment in the primary classes in the *t*-th year in the *i*-th district

$WTBB_t^i$ = Percentage of students in primary educational institutions without blackboard in the *t*-th year in the *i*-th district

= (Enrolment in primary educational institutions having no blackboard in the *t*-th year in the *i*-th district/Enrolment in primary educational institutions in the *t*-th year in the *i*-th district) × 100

$rept_{t-1}^i$ = Repetition rate in the primary classes in the (*t*–1)-th year in the *i*-th district

$$= \frac{REPT_t^i}{ENR_{t-1}^i} \times 100$$

= (Total number of repeaters in primary classes in the *t*-th year in the *i*-th district/Total enrolment in primary classes in the (*t*–1)-th year in the *i*-th district) × 100

It is assumed that there is a linear relationship between drp_t^i and the independent variables.

Therefore, the above functional form can be written as a linear equation, which is shown as follows:

$$drp_t^i = \beta_1 + \beta_2 SC_t^i + \beta_3 ST_t^i + \beta_4 SCR_t^i + \beta_5 GI_t^i + \beta_6 WTBB_t^i + \beta_7 rept_{t-1}^i + u_t^i \quad \ldots(2)$$

Where u_t^i is the stochastic error term for the t-th year and *i* stands for the individual district.

This study is based on the district-level secondary data which are available from various issues of District Report Cards of District Information System for Education (DISE) published by National University of Educational Planning and Administration (Arun C. Mehta, 2006 to 2010).

The economic logic behind the inclusion of the explanatory factors is as follows:

Firstly, 'percentage of SC enrolment in primary classes' and 'percentage of ST enrolment in primary classes' have been considered as two important explanatory factors for dropping out from the primary classes over the years. It is evident that Scheduled Caste and Scheduled Tribe people are the most vulnerable and poorest section of the society.

Often it is found that the survival of these poorest families depends on the children's earnings. Therefore, it can be assumed that even if the children of these poor backward class people are enrolled in the primary schools, the pressure of running the family may have compelled the parents to withdraw their children from the schools after few years and send them to work. Therefore, it can be assumed that high percentages of enrolment of Scheduled Caste and Scheduled Tribe students in the primary level would result in high drop-out rate in the primary classes.

Secondly, as a proxy variable for overcrowding in primary level classrooms, 'student-classroom ratio in primary classes' is considered as another important explanatory factor for dropping out from the primary classes over the years. If classrooms are not insufficient in the primary schools compared to the number of students, i.e., if the classrooms are overcrowded, it becomes nearly impossible for the teachers to take care of each and every student separately. Therefore, it can be supposed that in such a situation, lack of personal care would discourage the children to remain connected with the schools and would result in early drop-outs.

Thirdly, as a proxy variable for dominance of girl students in total enrolment in primary classes, 'Gender Parity Index in primary classes' has been considered as an important explanatory factor for drop-out from the primary classes in the districts of Tripura. It is assumed that the higher the GPI in primary classes is, the higher will be the drop-out rate in the same classes. It can be supposed that, education is still not considered necessary for girl children and they are generally kept at home to do household chores and are prepared for marriage. Therefore, even if the girls are enrolled in the primary schools, after a few years they drop out from there. It is further assumed that non-availability of enough number of separate schools for girls and non-availability of women teachers in the schools are also responsible for dropping out of girls from the primary classes.

Fourthly, 'percentage of students in primary educational institutions without blackboard' is another important explanatory factor for the drop-out in primary classes. The blackboard is probably the most convenient, effective and widely used basic visual teaching aid which is absolutely necessary in the learning process for the primary level students. Therefore, absence of a blackboard in a primary level school can be assumed to act as a significant obstacle to retain the students in the primary level classes.

Finally, 'repetition rate in the primary classes in the previous year' has been considered as one of the most important explanatory factors influencing drop-out in the primary level. Early grade repetition can assume to be a negative signal about a child's ability in the minds of the parents. Therefore, early grade repetition may be assumed to diminish parents' belief in their children's ability and thereby result in dropping out.

Econometric Model

To estimate the drop-out equation (2) of primary level students in the districts of Tripura quantitatively the panel data regression method has been applied. The logic behind the use of this model can be given as follows.

The purpose of the study is to find out the proximate explanatory factors behind the drop-out problem of the students in the primary level educational institutions of the districts[2] of Tripura for the past five years, i.e., for the period from 2005-06 to 2009-10. Panel data regression model seems to the ideal method of analysis for this study for two reasons. Firstly, pooling the data for several years increases the number of observations and, therefore, increases efficiency in estimation and power in hypothesis testing. Secondly, a panel data regression model has some fundamental advantages over a cross-section model, *viz.*, greater flexibility in modelling differences across districts (such as the possibility of controlling for 'district effects'). Under these circumstances this study tries to find out the proximate explanatory factors behind the drop-out problem of the students in the primary level educational institutions of the districts of Tripura for the period from 2005-06 to 2009-10 using the panel data regression model. District specific effects can be thought of in two different ways—fixed effects or random effects. In the Fixed Effects model, the individual effect is, which is taken to be exogenous and constant over *t*, and to vary across districts. In the Random Effects model, the district specific effect is modelled as an additional time invariant error term for each district, i.e., this model specifies as group specific disturbances. The Random Effects model assumes that the district specific random error is uncorrelated with the other explanatory variables. To choose between the Fixed Effects and the Random Effects model, the Hausman Test is done. The null hypothesis of Hausman Test is that the Fixed Effects and the Random Effects model estimators do not differ substantially. The value of the test statistic is quite high and significant at the 4 per cent level. Therefore, the null hypothesis is rejected which implies that the Random Effects model is not suitable and the Fixed Effects model is the appropriate one for this analysis.

III

Results of the Econometric Analysis

Now the hypotheses are substantiated as well as the factors responsible for the abovementioned changes in the nature of drop-out problem of students in the primary level educational institutions of the districts of Tripura during the period from 2005-06 to 2009-10 are analyzed on the basis of the econometric results. Results of the analysis are presented in Annexure. The results of the analysis can be interpreted as follows:

First of all, it can be found that model specification is statistically significant and all the explanatory variables together explain quite a significant proportion (61%) of cross-district variation in drop-out of primary level students at the inter-temporal level. Further, it follows from the analysis that SC^i_t has been found to have negative and highly significant coefficient. This implies that 'percentage of SC enrolment in primary classes' has negative and highly significant influence on the drop-out rate over the period from 2005-06 to 2009-10. This result, in fact, contradicts the hypothesis that high percentages of enrolment of Scheduled Castes would result in higher drop-out rates. On the other hand, the result implies that as the percentage of Scheduled Caste enrolment increases over the years, drop-out rate significantly decreases. In other words, according to the results, Scheduled Caste students have been found to be significantly disinclined towards dropping out from their classes over the years.

Secondly, according to the results, ST^i_t has been found to have negative but insignificant coefficient. This implies that 'percentage of ST enrolment in primary classes' has negative but insignificant influence on the drop-out rate over the period from 2005-06 to 2009-10. This implies that, although it was hypothesized that high percentages of enrolment of Scheduled Tribe would result in higher drop-out rates, the result actually shows that as the percentage of Scheduled Tribe enrolment increases over the years, drop-out rate decreases. However, this is also evident from the result that increase in the Scheduled Tribe enrolment does not significantly decrease drop-out rate from the primary classes.

Thirdly, the variable SCR^i_t has been found to have negative and highly significant coefficient. This implies that 'student-classroom ratio in primary classes' has negative and highly significant influence on the drop-out rate over the period from 2005-06 to 2009-10. This result also contradicts the hypothesis that higher student-classroom ratio in primary classes would, in fact, result in higher drop-out rates from those classes. In other words, although it was hypothesized that if the number of students in the primary level classrooms increases over time, lack of personal care would gradually discourage the children to remain connected with the schools and would result in higher rates of drop-out, but the result shows that, in fact increase in the number of students in the primary level classroom has created an encouraging atmosphere for the students to remain connected with the schools and has been found to have a dampening effect on drop-out rate in those classes.

Fourthly, according to the result, the variable GP^i_t has a positive and highly significant coefficient. It is also worthmentioning that the value of the coefficient of GPI^i_t is very high and in fact, is the highest among the coefficients of all the other explanatory factors. This result supports the hypothesis that the higher the GPI in primary classes is, the higher will be the drop-out rate in the same classes. In other words, it is evident from the result that the girl students have been the major part of the dropped out students during the period from 2005-06 to 2009-10. The result also implies that with the increase in the Gender Parity Index of enrolment in the primary classes, drop-out rate increases significantly.

Fifthly, the variable $WTBB^i_t$ has been found to have a positive and highly significant coefficient. This implies that 'percentage of students in primary educational institutions without blackboard' has positive and highly significant influence on the drop-out rate over the period from 2005-06 to 2009-10. This result supports the hypothesis that the higher the 'percentage of students in primary educational institutions without blackboard', the higher will be the drop-out rate in the primary classes. The result implies that the absence of blackboards in the primary level schools has been a significant obstacle to retain the students in the primary level classes and, therefore, has been a significant positive explanatory factor of drop-out of students in the primary classes.

Finally, from the result of the analysis, it is found that the variable $rept^{i}_{t-1}$ has positive and significant (at the 8% level) coefficient. This implies that the result has substantiated the hypothesis. The result shows that the higher the repetition rate in the previous year the higher will be the drop-out rate in the present year. In other words, the result substantiates our hypothesis that early grade repetition diminishes parents' belief in their children's ability and increases the drop-out rate significantly.

IV

Concluding Observations

This chapter is a humble attempt to find out the major issues and challenges of primary education in Tripura. Since the problem of drop-out turns out to be the major challenge, this study seeks to find out the correlates of drop-out of students from the primary level educational institutions in the districts of Tripura over the period from 2005-06 to 2009-10 in terms of Panel Data Regression Model and also to see the nature of relative contribution of the explanatory factors during the period of study. Drop-out function is estimated using the drop-out rate as the dependent variable and 'percentage of SC enrolment in primary classes', 'percentage of ST enrolment in primary classes', 'student-classroom ratio in primary classes', 'Gender Parity Index', 'percentage of students in primary educational institutions without blackboard' and 'repetition rate in the previous year' as independent variables. The econometric analysis reveals that 'percentage of SC enrolment in primary classes', 'student-classroom ratio in primary classes', 'Gender Parity Index' and 'percentage of students in primary educational institutions without blackboard' are the most significant explanatory factors throughout the whole period. The result also shows that 'repetition rate in the previous year' is a significant explanatory factor of drop-out during the period of study.

Major conclusions regarding the drop-out of students in the primary level educational institutions in the districts of Tripura during the period from 2005-06 to 2009-10 can be outlined as follows.

Firstly, during the last five years, i.e., from 2005-06 to 2009-10, the Scheduled Caste students in the primary level education institutions of Tripura, have been found to be significantly reluctant to drop-out from their classes. With the increase in the enrolment of Scheduled Caste students, drop-out rate has been found to be significantly declining.

Secondly, result of the study reveals that although the Scheduled Tribe students in the primary level schools of Tripura have been found to be unenthusiastic of dropping out from their classes but such unwillingness has not been significant enough to reduce the overall drop-out rate during the period of the study.

Thirdly, during the period of the analysis, increase in the number of students in the classrooms of primary level educational institutions has created an encouraging atmosphere for the students to remain connected with the schools and thereby has reduced the drop-out rate in these classes.

Fourthly, throughout the whole period of the study, girl students have been found to be the major part of the dropped out students in the primary level schools of Tripura. With the increase in the enrolment of girl students, drop-out rate has increased significantly.

Fifthly, the result of the analysis indicates that during the whole period, absence of blackboards in the classrooms of the primary level educational institutions has been a highly significant reason of drop-out of students from those classes.

Finally, according to the result, repetition in the previous year has been found to have a negative influence on the completion of primary level of education and has increased the drop-out rate in the current year in that level.

The major policy implications for the drop-out problem in the primary classes of Tripura which emerge from this study are as follows.

Firstly, according to the result, Scheduled Caste students in the districts of Tripura have been found to be highly enthusiastic for completing their primary level of education. To build up such enthusiasm, government should continue to encourage them through all kinds of concessions in education, free supply of text books, remedial teaching for weak students, merit scholarships, etc. Such schemes would not only reduce the drop-out rate in the primary classes but encourage the students to pursue their studies in the higher levels also.

Secondly, the analysis reveals that during the whole period of analysis, although Scheduled Tribe students have shown their interests in the completion of primary level education but such enthusiasm has not been strong enough to reduce the overall drop-out rate in the primary classes. To make Scheduled Tribe students more enthusiastic in completing their studies, government should spread awareness about the benefits of completing at least the primary level of education among the Scheduled Tribe people of the State. Moreover, government should provide all kinds of concession in primary education, free supply of text books, remedial teaching for weak students, merit scholarships, etc. for Scheduled Tribe students to retain them in the primary classes.

Thirdly, according to the result, increase in the number of students in the classrooms of the primary level schools have been found to have a demonstration effect on the students to complete their primary level education and thereby has created an enthusiastic atmosphere which has resulted in the reduction of drop-out rates in those classes. Under these circumstances, to reduce the drop-out rate, government should provide all the infrastructural facilities for quantitative as well as qualitative improvement of the classrooms in the primary level educational institutions of Tripura so that increasing number of students get attracted to join the classes and complete the primary level of education.

Fourthly, result of this study reveals that girl students have been the major part of the dropped out students in the primary level classes in the districts of Tripura during the period from 2005-06 to 2009-10. To reduce the overall drop-out rate in the primary classes, therefore, drop-out of the girl students should definitely be reduced. For this purpose, government should sensitize the parents of girl students about the necessity of primary level of education for the girls. Government should convince the parents that the girls should have the right to education just like the boys and they should not be kept at home to do household chores and be prepared for marriage. In addition to this, government should provide separate toilets for girls in all the primary level schools and increase the number of lady teachers in each school. In addition to this, enough number of separate school girls should be established to encourage them to complete their grades.

Fifthly, the analysis indicates that absence of blackboards in the classes of primary level schools has been a significant obstacle to retain the students in the schools of Tripura. Therefore, to reduce the overall drop-out rate, government should take necessary steps to provide blackboard in each and every classroom of the primary level schools in the districts of Tripura so that with the effective use of blackboards, the classrooms become live and active and thereby students become interested in sustaining their studies.

Finally, repetition in the previous year has been found to be a significant reason for dropping out from the primary classes in the present year, throughout the whole period of the study. To reduce repetition rate, government should focus on improving the quality of primary level schooling through proper teacher training and free distribution of textbooks. Government should also make nutritional and health interventions in the early childhood of the students to ensure normal cognitive development of these children to pre-school education. As a supportive strategy, government should educate the parents and provide public subsidies to the poor families to defray the direct and opportunity costs of primary education.

REFERENCES

Banik, A. and D. Neogi (2011). *Poverty and Earning Compulsions for the Family Pushed Tribal Children out of Schools: Dropout Study in Basic Education Level in Tribal Areas of North-East India, Social Science Research Network Working Paper 1762211*. New Delhi: Social Science Research Network

Cairns B.R., Cairns D.B. and J H. Neckerman (1989). Early School Dropout Configurations and Determinants. *Journal of Child Development*, Vol. 60(6), 1437-52.

Hausman, J.A. (1978). Specification Tests in Econometrics. *Econometrica*, Vol. 46, 1251-1271.

Jimerson, S.; Egeland, B. and L.A. Sroufe (2000). A Prospective longitudinal study of high school dropouts examines multiple predictors across development, *Journal of School Psychology*, 38(6), 525-549.

Jimerson, S.; Carlson, E.; Rotert, M.; Egeland, B. and L.A. Sourie (1997). A Prospective, Longitudinal Study of the Correlates and Consequences of Early Grade Retention, *Journal of School Psychology*, Vol. 35, 3-25.

Louise Ash. (2007). Dropout rate in India link to mother's education level, *The Hindustan Times*, 8th January 2001.

Mehta, Arun C. (2006). *Elementary Education in India : Where Do We Stand? DISE State Report Cards 2005-06*. New Delhi: NUEPA.

Mehta, Arun C. (2007). *Elementary Education in India : Where Do We Stand? DISE State Report Cards 2006-07*. New Delhi: NUEPA.

Mehta, Arun C. (2008). *Elementary Education in India : Where Do We Stand? DISE State Report Cards 2007-08*. New Delhi: NUEPA.

Mehta, Arun C. (2009). *Elementary Education in India : Where Do We Stand? DISE State Report Cards 2008-09*. New Delhi: NUEPA

Mehta, Arun C. (2010). *Elementary Education in India : Where Do We Stand? DISE State Report Cards 2009-10*. New Delhi: NUEPA

Rumberger, R.W. (1983). Dropping out of High School: The Influence of Race, Sex and Family Background, *American Educational Research Journal,* Vol. 20 (2), 199-220.

Rumberger, R.W. and A K. Larson (1998). Student Mobility and the Increased Risk of High School Dropout, *American Journal of Education*, Vol. 107(1), 1-35.

Zaidi, S.M.I.A. (2005). Provision and Utilization of Elementary Education Facilities in North-East India: What D.I.S.E. Data Reveal? *New Frontiers in Education*, Vol. 35(3).

Annexure

Pooled Regression (Ordinary Least Squares)

Ordinary least squares regression Weighting variable = none
Dep. var. = drp^{i}_{t} Mean = 7.925000000, S.D.= 4.659893267

Model size:	*Observations =20*	*Parameters = 7*	*Deg. Fr.=13*
Residuals:	Sum of squares = 129.4125237		Std. Dev.= 3.15512
Fit:	R-squared =.686332		Adjusted R-squared = .54156
Model test: F	F [6 13] =	4.74	Prob value = .00904
Diagnostic:	Log-L = -47.0515	Restricted (b=0)	Log-L = –58.6457

Log Amemiya Pr Crt.= 2.598, Akaike Info. Crt.= 5.405

Panel Data Analysis of DRP [ONE way]

Unconditional ANOVA (No regressors)

Source	*Variation*	*Deg. Free.*	*Mean Square*
Between	124.945	3	41.6485
Residual	287.632	16	17.9770
Total	412.578	19	21.7146

Variable	*Coefficient*	*Standard Error*	*t-ratio*	*|P[|T|>t]*	*Mean of X*
SC^{i}_{t}	– 3.486200859	.83001548	–4.200	0010	17.740000
ST^{i}_{t}	–.5996172298	.19090252	–3.141	0078	46.275000
SCR^{i}_{t}	–.8737524589	.31969080	–2.733	0171	20.500000
GPI^{i}_{t}	77.85622927	38.055706	2.046	0616	.92350000
$WTBB^{i}_{t}$	.4338527805	.11145414	3.893	0019	6.4600000
$rept^{i}_{t-1}$	.1634492065	.18405579	.888	.3907	9.2700000
Constant	39.21132511	47.182494	.831	.4209	

Pooling the data for five years has increased the number of observations and, therefore, has increased the efficiency in estimation and power in hypothesis testing. But this pooled regression model lacks the flexibility because we cannot control for the district-level effects. Therefore, we reject this pooled regression model.

Fixed Effect Model with Group Dummy Variables (Suitable for this Study)

Ordinary	*Least Squares Regression*	*Weighting variable = none*
Dep. var. = DRP	Mean = 7.925000000	S.D.= 4.659893267

Model size:	Observations = 20,	Parameters = 10,	Deg. Fr.= 10
Residuals	Sum of squares = 85.46089747	Std. Dev.= 2.92337	
Fit:	R-squared = .792861,	Adjusted R-squared =.60644	
Model test:	F[9, 10] = 4.25,	Prob value =.01688	
Diagnostic:	Log-L = –42.9020,	Restricted (b = 0)	Log-L = –58.6457

Log Amemiya Pr Crt. = 2.551, Akaike Info. Crt.= 5.290
Estd. Autocorrelation of $e(i, t)$ .000000

Variable Coefficient	*Standard Error*	*|t-ratio |*	*P*	*[|T|>t] |*	*Mean of X*
SC^i_t	–4.379591788	1.9156660	–2.286	.0383	17.740000
ST^i_t	–1.763709527	1.0811106	–1.631	.1251	46.275000
SCR^i_t	–1.093454835	.33029479	–3.311	.0052	20.500000
GO^i_t	91.31711091	44.507572	2.052	.0594	.92350000
$WTBB^i_t$	.3292104601	.11887195	2.769	.0151	6.4600000
$rept^i_{t-1}$	.4671750071	.25365797	1.842	.0868	9.2700000

Random Effects Model: $v(i, t) = e(i, t) + u(i)$
Estimates: Var[e] = .854609D + 01
Var[u] = .147369D + 01
Corr[$v(i, t)$, v (i, s)] = .147078
Fixed *vs*. Random Effects (Hausman) = 13.38
(6 df, prob value = .04)
(High (low) values of H favour FEM (REM).)

Reestimated using GLS coefficients:

Estimates: Var[e] = .106806D + 02
Var[u] = .285984D + 01
Sum of Squares = .131786D + 03
R-squared = .680578D + 00

Variable	*Coefficient*	*Standard Error*	*|b/St.Er.|*	*P[|Z|>z]*	*Mean of X*
SC^{i}_{t}	–3.533730426	.86316957	–4.094	.0000	17.740000
ST^{i}_{t}	–.6233813305	.19543222	–3.190	.0014	46.275000
SCR^{i}_{t}	–.9381620551	.30311357	–3.095	.0020	20.500000
GPI^{i}_{t}	72.29395912	36.597803	1.975	.0482	.92350000
$WTBB^{i}_{t}$	.4182346526	.10450529	4.002	.0001	6.4600000
$rept^{i}_{t-1}$	.2213158094	.18384143	1.204	.2287	9.2700000
Constant	47.17580651	47.783748	.987	.3235	—

A panel data regression model has some fundamental advantages over a cross-section model, *viz.*, greater flexibility in modelling differences across districts, *i.e.*, the district-specific effects. District specific effects can be thought of in two different ways—fixed effects or random effects. In the Fixed Effects model, the individual effect is, which is taken to be exogenous and constant over time *t*, and to vary across districts. In the Random Effects model, the district specific effect is modelled as an additional time invariant error term for each district, *i.e.*, this model specifies as group specific disturbances. The Random Effects model assumes that the district specific random error is uncorrelated with the other explanatory variables. To choose between the Fixed Effects and the Random Effects model, the Hausman Test is done. The null hypothesis of Hausman Test is that the Fixed Effects and the Random Effects model estimators do not differ substantially. The value of the test statistic is quite high and significant at the 4 per cent level. Therefore, the null hypothesis is rejected which implies that the Random Effects model is not suitable and the Fixed Effects model is the appropriate one for this analysis.

10

Looking at Nagaland through MDG Lens: An Appraisal

Ratna Roy Sanyal and *Subhashree Sanyal*

Nagaland is a vibrant hill State located in the extreme North-Eastern end of India, bounded by Myanmar in the east; Assam in the west; Arunachal Pradesh and a part of Assam in the north with Manipur in the south. It offers rich incomparable traditional and cultural heritage. According to the Census of 2011, Nagaland has a total population of 1,980,602 people with 1,025,707 males and 954,895 females. The sex ratio is 931 and the density of population is 119 per square kilometre. Nagaland has also recorded a decadal growth rate of minus 0.47 which is a record in the history of this country. The State consists of seven Administrative Districts, inhabited by 16 major tribes along with other sub-tribes. Each tribe is distinct in character from the other in terms of customs, language and dress.

The economy of Nagaland mostly consists of Agriculturalist and around 75 per cent of the population lives in the rural areas. About one-third of this rural population can be considered to be below the poverty line. Among the people living in urban areas one-fifth of them are below the poverty line. The economy of Nagaland is primarily forest-based and agricultural. It has very rich forest resources. Nagaland is also rich in mineral resources including coal, limestone, iron, nickel, cobalt, chromium, and marble. The approximate reserve of limestone is 1000 million tonnes and substantial reserves of marble and decorative stone, petroleum and natural gas, nickel, cobalt and chromium are available in the State.

But these mineral resources are yet to be explored and exploited. About 68 per cent of working population is dependent on agriculture. The first Human Development Report was published for Nagaland in 2004 and it has been seen that Nagaland has fared well compared to other States in the country. The HDI of the State was estimated to be as high as 0.62 compared to the national average of 0.472 in 2001. The State's high GDP, Education and Health levels contribute to this estimate. A glance at Nagaland from the perspectives of MDG would help analyze where Nagaland stands in India's development map and the current overall status of progress of the State.

Goal 1: Eradication of Extreme Poverty and Hunger

New official estimates for 2009-10 released by the Planning Commission of India (PCI) revealed that poverty level in Nagaland saw the highest rise among the other States in the country. As per the estimates, Nagaland saw the highest rise of 12.1 percentage points in poverty between 2004-05 and 2009-10, followed by Manipur (9.2 percentage points), Mizoram (5.7 percentage points), Assam (3.5 percentage points) and Meghalaya (1 percentage point).

It further suggests that Nagaland has 20.9 per cent (4.1 lakh) population living below poverty line in 2009-10 respectively, as against 37.9 per cent (9 lakh), 15.4 per cent (4.1 lakh) and 8.8 per cent (1.7 lakh) population in 2004-05 respectively.

Nagaland has its rich tribal culture, a lot of focus on indigenous livelihood, self-employment generation programmes, handicrafts will help improve the situation of the population in the State. Nagaland has a suitable soil condition for bamboo and with little interest and care given to the "wonder plant"—bamboo, it would be enough to eliminate poverty in the society. Bamboo cultivation and its products would help tap vast resources useful for the Naga population specially generating their own livelihood. The State government has been investing in a lot of income generation activities to combat the situations of unemployment and poverty.

Goal 2: Achieve Universal Primary Education

Literacy rate in Nagaland has seen upward trend and is 80.11 per cent as per 2011 population census. Of that, male literacy stands at 83.29 per cent while female literacy is at 76.69 per cent. In 2001, literacy rate in Nagaland stood at 66.59 per cent of which male and female were 76.04 per cent and 56.87 per cent literate respectively. In actual numbers, total literates in Nagaland stands at 1,357,579 of which males were 731,796 and females were 625,783. Average Literacy rate in Nagaland for Urban regions was 90.21 per cent in which males were 92.11 per cent literate while female literacy stood at 88.10 per cent. Total literates in urban region of Nagaland were 452,780. In rural areas of Nagaland, literacy rate for males and female stood at 79.49 per cent and 72.01 per cent. Average literacy rate in Nagaland for rural areas was 75.86 per cent. Total literates in rural areas were 904,799. The total number of literates has gone up from 1,132,323 in 2001 to 1,357,579 in 2011 [Census, 2011].

At present there is one Central University and a number of private colleges and State universities facilitating higher education across the State. The total budget of the Department of Higher Education during 2009-10 is Rs.3862.48 lakhs. In Nagaland, there are 51 colleges, out of which 38 are under private management. Out of about 29,000 college students enrolled during 2009-10. The State with grants, scholarships and different schemes is making formidable progress in this direction.

Goal 3: Promote Gender Equity and Empowerment

Nagaland has consistently maintained a GDI of 0.7 and above in comparison to the national average of 0.6. This is mainly because of the Naga culture of equal care for the girl child. Naga tribal values and practices have provided a unique framework of gender relations and access to health and nutrition. The contributing factors to the notable GDI in Nagaland are education and the improving opportunity for livelihood. The population of females is 9,54,895 compared to 10,25,707.

The Gender Development Index of Nagaland is 0.420 as against 0.623 for HDI. The Nagaland State Human Development Report highlights the inter-district disparity in human development attainments. Among the eight districts in Nagaland, Dimapur with a HDI value of 0.733 ranks highest whereas Mon with a HDI value of 0.450 ranks the lowest. With regard to the gender related development index (GDI), Kohima ranks the highest whereas Mon has the least GDI value. For the Human Poverty Index, while Mokochung has the lowest value, Mon again has the highest levels of human poverty. In fact, the districts of Mon and Tuensang are consistently ranked eighth and seventh on all the three indices, highlighting the need for concerted strategies for these districts. The cultural implications impact on the gender equity factor of Nagaland which is different from rest of the country.

Goal 4: Reduce Child Mortality

Health indicators in Nagaland revealed that Central and State government expenditure on health has increased to 1.4 per cent of GDP in 2011-12 from less than 1 per cent at the beginning of the Eleventh Plan. Nagaland fare better when compared with the national average. It is interesting to note that in Nagaland, in spite of the fact that the percentages of pregnant mothers who go for antenatal check-up, mothers who deliver babies in health centres/hospitals, and children, who are fully immunized, are all poor; Infant Mortality Rate (IMR) is two times lower than the national figure. IMR for Nagaland is 23 (Number of infant deaths per 1000 live births) against the national average of 47 as per Sample Registration System bulletin, December 2011.

The National Family Health Survey third round (NFHS III) which was carried out in 2005-06, it was observed that in Nagaland, children deaths are related to the educational status of the mothers. From the study it has been seen that children die due to lack of education of mothers for pregnancy care, child birth care and no idea of spacing in child birth. Women sometimes lack elementary knowledge

Though there is awareness about vaccination and immunization programmes in Nagaland, lack of concern for health care among the population becomes the major cause of infant deaths.

Goal 5: Improve Maternal Mortality

Health Infrastructure and skilled personnel play a very important role in ensuring availability of Health Services in any region. The percentage of people who received anti-natal care is 1.9. There has been 11.6 per cent births through safe institutional delivery in Nagaland and is slowly improving.

Cultural factors play an important role, the idea of matriarchal lineage, Christianity and prevailing western culture especially in the tribal dominated States helps the overall the North-Eastern Region fares well in terms of sex-ratios, lower maternal mortality and fewer teenage pregnancies. Use of preventive methods for family planning is poorly institutionalized in this region. As per the recent Census data of 2011 0.49 per cent married women use contraceptives which are progressive and still improving. Therefore, maternal mortality results due to lack of care of pregnant mothers especially in rural areas of the State.

Goal 6: Combat HIV/AIDS Malaria and Other Diseases

The battle against HIV/AIDS in the tiny State of Nagaland is far from over as though the prevalence rate has officially dropped to 0.78 per cent since December 2010 from that of 1.27 per cent till October 2010, disturbing trends have emerged that suggest that the number of people infected with HIV may actually be on the rise

The total number of persons detected with HIV/AIDS in Nagaland till December 31, 2011, stands at 12,440 with Dimapur district accounting more than 4,000 cases. The total number of unregistered HIV/AIDS cases is anybody's guess as stigma and discrimination against persons infected with HIV/AIDS still continues. And while HIV infection through drug injection has climbed down to 3.5 per cent, transmission through sexual route has gone up to 89 per cent in Nagaland. Efforts are being made by the State Aids Control Society for promoting safe sex practices and also reducing usage of syringes by drug users.

Goal 7: Ensure Environment Sustainability

The topography of Nagaland is majorly hilly. According to Environment Sustainability Index, the State stands 4 on ESI Scale. The impact on Human Health and Disaster is 0.60 which is much lower compared to other States. The land-use and agriculture stands on an average of 0.69 on ESI Scale. Report states that Nagaland stands highest on the ESI Scale as of 2011.

The greater part of Nagaland is hilly and rugged terrain. Though the vegetation cover is dwindling, still there are many areas which are inaccessible. The stretch of land bordering Assam is flat and low lying. Anthropogenic activities which include *jhum* cultivation, deforestation and reckless burning of forest land are a matter of concern for environmental degradation and soil erosion. Natural damage to the environment is also a major factor of concern as the land is hilly and rugged, and the adverse climatic conditions of very cold winters, hot summers and heavy rains have very serious bearings on the soil.

There are different hazards which Nagaland is prone to, mainly Earthquake, Landslide and Cloudburst. Efforts should be made by the State to construct their structures [both government and private] with expertise. Land-use need to be planned to prevent landslides and flood. 30 per cent area of Nagaland is known as rice bowl State.Commerical crops include sugarcane, jute and potato.

Goal 8: Develop a Global Partnership for Development

Nagaland is predominantly governed by its indigenous culture. There is a need to bring holistic development across the region. Having a rich resource base the State can promote tourism and open up gates for investments in this sector. The uniqueness of culture can be a bargain point for global partnerships and development. Development of infrastructure, roads, easy communication services, health care, sanitation and ensuring safe drinking water is the need of the hour. Therefore State with the partnership of Civil Society Organizations and other development authorities like DoNER and NEC can bring about global partnership and progress in the State. More encouragement towards promotion of local culture, livelihood generation and unique products which can be an attraction for global partnerships and also in turn bring income generation to the State.

Analysis and Conclusion

Nagaland has made impressive strides in the last few decades, particularly, in areas such as attainments in literacy, health related indicators, health care services, water and sanitation, power, governance, women in development, and fiscal reforms. The unique initiative of 'Communitisation' of public institutions and services, in the sectors of education, health care and power distribution has enhanced accountability and involvement of both the community and the Government towards greater effectiveness. Enormous potential exists in the State for eco-tourism. It is time when the strengths, richness and abundance of natural and human resources and the social capital of Nagaland are harnessed. There is an air of hope and expectation among the people of Nagaland. The need is to change the mindset, which generally looks at the various structural handicaps of Nagaland, like its remoteness and backwardness, as blocks or limitations, and begin to view them as challenges and potentials waiting to be channelled and realised. (Government of Nagaland, 2004) Nagaland is blessed with abundant natural resource, mixes cultivation and a very strong indigenous culture. The State is lagging behind in terms development and infrastructure primarily. Though the government is focusing on development of North Eastern Region there is a special need to explore need based intervention in the State. All government programmes and policies implemented in the State need fruitful implementation. Expertise to build roads, housing should be brought into the state. The State is already ahead in terms of health and education compared to others in North-Eastern India and this can be further accentuated to achieve a higher HDI. Developmental gaps needs to be bridged which can be done through the three sector partnerships—government, corporate and civil societies. Indigenous culture and products need to be promoted. Youth may be involved to act as catalysts and agents to social change. A holistic developmental approach is the need of the hour to promote a better HDI, invite tourism and accentuate the achievement of millennium commitments.

REFERENCES

Government of Nagaland (2004). *Nagaland State Human Development Report*. Kohima: Department of Planning & Coordination

Ganguly, J.B. (1995). *Urbanization and Development in North-East India*. New Delhi: Deep & Deep Publication

North-East Council (2002). *Basic Statistics of North-East Region*. Shillong: NEC

Sanyal, S. (2012). Looking at Nagaland through MDG Lens. *Yojana,* Vol. 60.

Yonuo, A. (1974). *The Rising Nagas: A Historical, Political Study*. New Delhi: Vivek Publishing House

11

Socio-Economic Development of Arunachal Pradesh: A Study-based Reflections from Nyishi Tribal Village

Tame Ramya

Introduction

This chapter examines the some aspects of development in Arunachal Pradesh and impact of developmental efforts made by the government and non-government agencies to bring required socio-economic changes in the lives of marginalized tribal communities in Arunachal Pradesh. The analysis is made in the context of changes that have taken place in the very concept of development and the shifts of responsibility by the welfare State. Till the emergence of new economic policy of Government of India and consequently liberalization of economic policies it is felt that the responsibilities of a welfare state must be not only as regulator but also provider of infrastructure so that the poor get freed themselves from the cycle of poverty (Gupta, 2004:1-5). As a part of structural adjustment they had to adapt new economic policies in which the State had to shake off its primary responsibilities of providing welfare measures to the poor. Meanwhile, the meaning of the concept of development also changed. It is in this context that the current paper tries to understand the impact of developmental efforts made by central and State government departments on Nyishi, numerically the largest tribe, in a village called Hiya in Kurung Kumey district of Arunachal Pradesh.

This chapter has been organised into five sections: The first section briefly explains the concept of development; the second section deals with the brief aspects of development processes in Arunachal Pradesh; the third section deals with profile of the Nyishi tribal village, i.e., area of study, their livelihoods; the fourth section looks at the development interventions of governmental and non-governmental agencies and final section illustrates the conclusions.

Concept of Development

The term development has been used in a wider sense. The purpose of development is to provide increasing opportunities to all people for a better life. It is essential to bring about more equitable distribution of income and wealth for promoting social justice and efficiency of production, to provide a greater variety of facilities like education, health services, nutrition, housing, etc. The developmental parameters of each tribe are different and it depends on the inhabitants and their settling conditions. Development interventions have varying impacts on the lives of poor people, ranging from the beneficial to the harmful or sometimes with no effect at all. However, it is always a challenge for researchers to assess the true effects of development interventions and to form a balanced view of overall patterns, especially over the long term.

According to International Labour Organisation (ILO), *Development involves 'humans' as distinct from material product. It is defined as a process which involves improvement in the quality of life of weaker sections and a greater participation and involvement of the masses in the process of decision making in the economic, social, political and cultural life of a society.*

According to Denis Goulet[1] (2010), *Development is not a cluster of benefits given to the people in need, rather a process by which a populace acquires a greater mastery over its own destiny.* Schumpeter (2002) defines development as only such changes in economic life that are not forced upon it from without, but arise by its own initiative from within.

According to Dudley Seers (1969), *Development means creating condition for the realisation of human potential.* Development is an elusive concept and evolves mobilisation of natural resources, augmentation of trained manpower, capital and technical know-how and their utilisation for attainment of constantly multiplying national goals, higher living standards and the change over from a traditional to a modern society. The essence of development is generally perceived as industrialisation and modernisation. Development is a multi-dimensional and multi-linear process.

Development is usually conceived as an aspect of change that is desirable, broadly planned and administered or at least influenced by governmental action. Thus the concept of development consists of (a) an aspect of change; (b) a plan or prediction; and (c) involvement of the government for the achievement of that planned or predicted goal. The term development is also used for the process of allowing and encouraging people to meet their own aspiration. The main aim of development is to increase national as well as per capita income and to raise the standard of living of the people and secure justice, freedom, equality and security for them in society. The focus of development is now increasingly on (a) Equitable distribution of wealth and income; (b) Full utilisations of manpower; (c) Better utilisation of natural resources; and (d) Protection of human environment, etc. Hence, Development means change plus growth, i.e., it includes growth, modernisation, increase in social facilities, etc.

Arunachal Pradesh: Brief Overview

Arunachal Pradesh, the twenty-fourth State of Indian Union, and situated in the easternmost corner of North-East India is a heterogeneous tribal State characterised by extraordinary ethnic, cultural, religious and linguistic diversity. This predominantly hilly State with an area of 83,743 sq.km is the largest State among the North-Eastern States of the country. Area-wise Arunachal Pradesh is 32.83 per cent of the North-East and around 6.76 per cent of India as a whole. It forms a complex hill system of Shivalik and Himalayan origin and is criss-crossed by numerous rivers and streams. The State shares a total of 1,630 km of international boundary with neighbouring countries; 1,030 km with China, 160 km with Bhutan and 440 km with Myanmar. The McMohan line defines the international boundary between India and China. Administratively, the State is divided into seventeen districts. Capital of the State is Itanagar in Papum Pare district. This sparsely populated State has a very low density of

17 people per sq km. Arunachal Pradesh is a fascinating State which is inhabited by 26 major tribes and more than 100 sub-tribes, each having their distinct culture, dialects and customs.

Prior to the attainment of Statehood in 1987, Arunachal Pradesh was known as the North-Eastern Frontier Tracts. It was in 1954 that it was renamed as North Eastern Frontier Agency (NEFA) and it was finally in 1972 that NEFA was given the status of a Union Territory. The constitutional development in the State provided the administrative mechanism for achieving sustained development. The initiation of the process of development in the State is thus a recent one, having unfolded in the years following independence of the country in 1947. Prior to that, the area was administered indirectly with the people enjoying a fair degree of autonomy in their villages and localities. The village or a group of villages bound by kinships/clanship/community enjoyed a considerable degree of autonomy.

In Arunachal Pradesh modernization is a post-independence phenomenon, beginning in slow but assured steps with the establishment of direct administrative structure in the State (Luthra, 1993). The establishment of institutional infrastructure in a hilly and sparsely populated region with a lot of inter-tribal differences was indeed a gigantic task. The problem was aggravated due to the operation of Inner Line Permit (ILP), which restricted the entry of outsiders into the State. The ILP was enacted by the British India Government in 1873 and continues to be in force. As per the provisions of this Act, people from other parts of the country cannot enter the State without the permission of the government. They also cannot own any fixed assets in the State. Further the induction of the cash economy in the region released the forces of institutional dynamics, which resulted in far-reaching changes and thus integrated the hitherto insular economics with the national economy. The most visible changes came into the institution of ownership, especially land.

Development Process in Arunachal Pradesh

While a comprehensive analysis of the development process in Arunachal Pradesh is away from the range of this chapter, some facets of it can be selectively looked at to have a better understanding of the challenges before the State and its people.

(i) Along with monetisation of the exchange process and gradual commercialisation of the economy, interpersonal inequalities in the distribution of income, assets and opportunities have developed manifold. While the elaborate networks of redistribution and reciprocity prevented the emergence of inequality in the traditional economy of the communities, under the mutually reinforcing influences of the market and the State, the gradual economic differentiation of the indigenous population has led to the emergence of a neo-rich, upwardly mobile class in Arunachal Pradesh. Occupationally this group belongs to the trading and business community, contractors or government services – but the most important aspect of their emergence is their access to the resources of the State.

(ii) The spatially uneven process of development in the State has created new challenges and constraints. Typically, the districts or areas bordering Assam, and within the districts areas near the urban, administrative centres have better infrastructural facilities than the interiors. Arunachal Pradesh has been the home of a number of tribes and sub-tribes, and many of them had limited mobility beyond their well-defined local boundaries. Inter-regional disparities in the State, therefore, has an additional implication – in many cases, up to an extent, it may reflect inter-tribal disparities.

(iii) Although Arunachal Pradesh until recently has been relatively free from secessionist violence and insurgency, its development performance, particularly in terms of the human development indicators, has not been satisfactory. Among the eight North-Eastern States, Arunachal Pradesh occupies fifth position in terms of Human Development Index (Table 11.1). In 2001, Arunachal Pradesh had the lowest literacy rate among all the North-Eastern States, as well as the highest urban-rural gap and the highest gender gap in literacy. In terms of enrolment, drop-out rates and school availability the States' performance, in a comparative perspective, has hardly been satisfactory (Table 11.2). In terms of health status of the population, Arunachal Pradesh has the second highest Under Five Mortality Rate and the third highest IMR among the North-Eastern States (Table 11.3).

Table 11.1: Human Development Index: Arunachal Pradesh and Other North-Eastern States

States	*Human Development Index*	
	1991	*2005*
Arunachal Pradesh	0.328	0.617
Assam	0.348	0.534
Manipur	0.536	0.707
Meghalaya	0.548	0.585
Mizoram	0.486	0.790
Nagaland	0.425	0.770
Sikkim	0.389	0.684
Tripura	0.381	0.608

Source: Website of DoNER

Table 11.2: Educational Attainments in Arunachal Pradesh and Other North-Eastern States

States	*Literacy Rates (in %)*	
	2001	*2011*
Arunachal Pradesh	54.74	66.95
Assam	64.28	73.18
Manipur	68.87	79.85
Meghalaya	63.31	75.48
Mizoram	88.49	91.58
Nagaland	67.11	80.11
Sikkim	69.68	82.20
Tripura	73.66	87.75

Source: National Human Development Report 2001.

Table 11.3: Health Status of the Population: Arunachal Pradesh and Other North-Eastern States

States	*Infant Mortality Rate*	*Under 5 Mortality Rate*	*% of Women with Anaemia*
Arunachal Pradesh	63.1	98.1	62.5
Assam	69.5	89.5	69.7
Manipur	37.0	56.1	28.9
Meghalaya	89.0	122.0	63.3
Mizoram	37.0	54.7	48.0
Nagaland	42.1	63.8	38.4
Sikkim	43.9	71.0	59.4
Tripura	–	–	59.4
INDIA	67.6	94.9	51.8

Source: National Family Household Survey-II (1998-99).

Table 11.4: Composition of Public Expenditure in Arunachal Pradesh and Other North-Eastern States during 1998-99

States	*Public Expenditure Ratio*	*Develop-ment Expenditure Ratio*	*Social Sector Expen-diture Ratio*	*Education Expenditure Ratio*	*Health Expenditure Ratio*
Arunachal Pradesh	65.09	72.98	26.27	12.04	5.43
Assam	˙20.50	62.39	39.07	26.34	4.65
Manipur	42.84	63.37	32.95	18.52	4.67
Meghalaya	33.63	69.10	35.19	16.95	7.22
Mizoram	–	71.16	35.35	12.97	4.93
Nagaland	–	53.89	25.34	9.55	5.39
Sikkim	208.40	28.13	14.99	7.31	2.84
Tripura	44.52	62.50	35.96	17.23	4.69
Central Government	43.70	34.90	8.20	3.90	1.80

Source: Planning Commission of India (2002).

While it is always possible to explain away this relative under-performance of Arunachal Pradesh by citing many historical, geographical and economic constraints, any forward looking strategy of development for the State needs to address these issues in their entirety as well as local specificities.

The Developmental Challenges in Arunachal Pradesh

Governance

Most of the districts in Arunachal Pradesh are backward, lack adequate power supply, education facilities, health care, livelihood avenues, and surface communication. Important border districts like Tawang suffer from bad road connectivity, a tenuous and risky helicopter service, poor electricity, and absence of good health care. The Arunachal Pradesh Human Development Report, 2005 identified infrastructure development as one of the key concerns of the State. This was followed up with Prime Minister Manmohan Singh's package of Rs. 24,000 crore in 2009 for building the 1,500 km Trans-Arunachal Pradesh highway. Yet, ground realities continue to be grim with the Border Roads Organisation (BRO) stating that it is an arduous task to build two-lane highways in such a hilly and inaccessible terrain.

Table 11.5: Funds Allocated to Education, Health and Infrastructure in Arunachal Pradesh: 2010-2011

(*in Rs.*)

Education	*Health and Family Welfare*	*Roads and Bridges*	*Forest*	*Hydro-Power Development*
394,52,08,000	156,76, 47,000	232,16,22,000	91,73,64,000	40,05,27,000

Source: *The Arunachal Pradesh Gazette Extraordinary*, Government of Arunachal Pradesh, March 29, 2010 at *http://india.gov.in/allimpfrms/allacts/3032.pdf* (Accessed on June 5, 2012).

The rural development schemes that Arunachal Pradesh has activated are Community Development Programme (COMM), National Rural Employment Guarantee Act (NREGA), Pradhan Mantri Gram Sadak Yojana (PMGSY), Rural Health (RH), Swarnjayanti Gram Swarozgar Yojana (SGSY), Multi-Sector Rural Development Programme (MSRDP), Providing Urban Amenities to Rural Areas (PURA), Below Poverty Line Schemes (BPLS), etc. (See Table 11.2).

As Arunachal Pradesh is 90 per cent dependent on the Union government for financial resources, the lacuna in these schemes is that they do not cater to local needs specifically.

Most schemes are too general in their focus given that they are formulated by the Central government. Therefore, it is the need of the hour that rural schemes should be formulated at the local level and attuned to the needs of the people. Significantly, the Ministry for the Development of the North-Eastern Region (DoNER) has given priority to Arunachal Pradesh and sanctioned 29 infrastructure projects amounting to Rs. 325.24 crore from Non Lapsable Central Pool of Resources (NLCPR) Fund in 2010-2011. The North Eastern Council (NEC) also allocated Rs 70.07 crore in 2011 for the same purpose[2].

Despite that, there is a noticeable tendency amongst the local population to compare the infrastructure development in other States to that of Arunachal Pradesh. In terms of tourism sector, Arunachal ranked last in the tourism sector amongst the North-Eastern States with Meghalaya taking the top spot[3]. In 2010, India invested around Rs. 5,944 crore (approximately $1.3 billion) under the consolidated fund in Arunachal Pradesh[4]. Significantly, the GDP growth rate of Arunachal Pradesh in 2010 was 0.12 per cent. While the State has

Table 11.6: District and Scheme-wise Release Position of Funds for Arunachal Pradesh during the Financial Year 2011-2012

(Rs. in lakhs)

Name of The District	*COMM*	*NREGA*	*PMGSY*	*RH*	*SGSY*	*DRDA*	*MSRDP*	*PURA*	*TRG*	*BPLS*	*TOTAL (Rs)*
Tawang	0.00	18.76	0.00	0.00	16.89	30.90	0.00	0.00	0.00	0.00	**66.55**
West Kameng	0.00	0.00	0.00	0.00	23.92	24.93	0.00	0.00	0.00	0.00	**48.85**
East Kameng	0.00	0.00	0.00	0.00	0.00	30.89	0.00	0.00	0.00	0.00	**30.89**
Papum Pare	0.00	696.28	0.00	0.00	15.82	24.93	0.00	0.00	0.00	0.00	**737.03**
Lower Subansiri	0.00	0.00	0.00	0.00	20.25	24.93	0.00	0.00	0.00	0.00	**45.18**
Kurung Kumey	0.00	0.00	0.00	0.00	0.00	30.90	0.00	0.00	0.00	0.00	**30.90**
Upper Subansiri	0.00	0.00	0.00	0.00	33.74	30.90	0.00	0.00	0.00	0.00	**64.64**
West Siang	0.00	0.00	0.00	0.00	0.00	35.23	0.00	0.00	0.00	0.00	**35.23**
East Siang	0.00	54.20	0.00	0.00	0.00	30.89	0.70	0.00	0.00	0.00	**85.79**
Upper Siang	0.00	0.00	0.00	0.00	0.00	30.90	0.00	0.00	0.00	0.00	**30.90**
Dibang Valley	0.00	0.00	0.00	0.00	12.63	24.93	0.00	0.00	0.00	0.00	**37.56**
Lower D/Valley	0.00	0.00	0.00	0.00	0.00	24.93	0.00	0.00	0.00	0.00	**24.93**
Lohit	0.00	4.17	0.00	0.00	36.99	24.93	0.00	0.00	0.00	0.00	**66.09**
Anjaw	0.00	0.00	0.00	0.00	0.00	24.93	0.00	0.00	0.00	0.00	**24.93**
Changlang	0.00	91.73	0.00	0.00	19.36	30.89	0.00	0.00	0.00	0.00	**141.98**
Tirap	0.00	0.00	0.00	0.00	0.00	30.90	0.00	0.00	0.00	0.00	**30.90**
Longding*	–	–	–	–	–	–	–	–	–	–	–
Total	**0.00**	**865.14**	**0.00**	**0.00**	**179.60**	**456.91**	**0.70**	**0.00**	**0.00**	**0.00**	**1502.35**

Source: Department of Rural Development.

*The data for Longding, the 17th district of the State is not yet available. The data for this district is included with Tirap district.

developed plans to expand the tourist industry, this is not supported by good infrastructure.

Table 11.7: Tourist Inflow into Arunachal Pradesh (2006-2008)

Tourist	*2006*	*2007*	*2008*	*CAGR (2004 base year)*
Domestic	80137	91100	89292	22.47%
International	706	2212	3020	50%-75%
Total	**80843**	**93312**	**92392**	**22.5**

Source: Tourism Development of Arunachal Pradesh, Ministry of Tourism, Government of India, February 2010 at http://tourism.gov.in/survey/Arunachal Pradesh-Feb-10.pdf (Accessed on June 5, 2012).

Riparian Issues

Riparian issues are significant in Arunachal Pradesh. Most of the rivers of the State have been identified as possessing enormous potential for hydro-power to the tune of 49,126 megawatt (MW) of electricity[5]. Arunachal Pradesh has signed nearly 103 memorandums of understanding (MoUs) with private groups to construct around 168 dams in the State. The State has five major river basins, namely, Kameng River Basin, Subansiri River Basin, Siang River Basin, Dibang River Basin, and Lohit River Basin on which 89 major hydro-electric projects are being planned as on date (See Figure 11.1). The largest river is the Siang, which is called the Yarlung Tsangpo in Tibet and the Brahmaputra in Assam. The Brahmaputra flows for around 5,80,000 square km across Tibet, Bhutan, India and Bangladesh. In India, its basin is shared by Assam (36.33%), Arunachal Pradesh (41.88%), Nagaland (5.57%), Sikkim (3.75%), West Bengal (6.47%) and Meghalaya (6.10%).

Private companies like the Jindal Group, Jaypee Arunachal Power Construction and Reliance have signed MoUs with the Arunachal Pradesh government to develop multipurpose hydro-electric projects believed to be largest power projects in the country. In 2006, an agreement was signed with the North Eastern Electric Power Corporation (NEEPCO) for development of about 10,230 MW of hydro power. The two most prominent projects are the 2,000 MW Lower Subansiri Project, the 405 MW Ranganadi Project and the 2,700 MW East Siang Project. Others include the Bharali Lift Dam

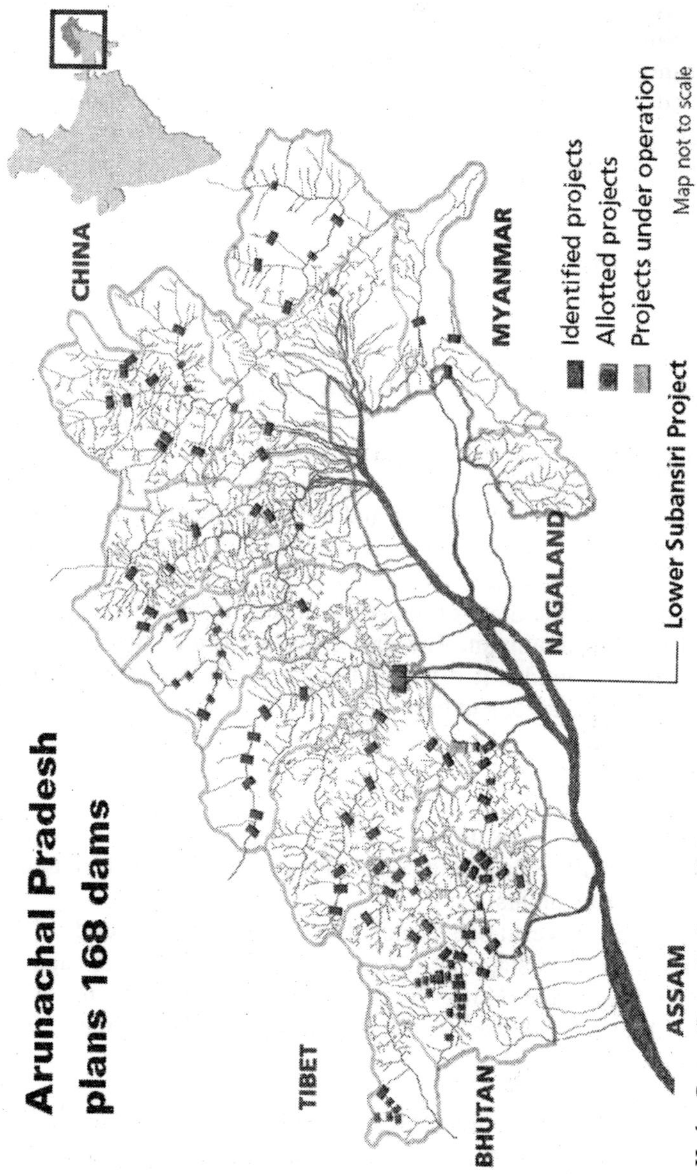

Fig. 11.1: Hydro-Power Dams in Arunachal Pradesh
Source: Department of Hydropower, Government of Arunachal Pradesh, Itanagar.

II on the Kameng River Basin with a capacity of 330 MW, the Hegio on the Subansiri with a projected capacity of 250 MW, etc. Significantly, local anxieties in Arunachal Pradesh are growing with regard to dams being built by private companies like Jaypee Arunachal Power Construction and Reliance Group. These fears are most apparent with regard to the East Siang River project being built by Jaypee.

Border Security

Border security is another major issue in Arunachal Pradesh. The shadow of the 1962 Indo-China war still looms large in the local political discourse. At present, the border roads are in bad condition and in some cases non-existent near the LAC in the places like Mechuka and Sarli area, and landslides obstruct smooth flow of traffic on the Along-Tuting-Gelling and Tawang-Bumla roadways. Most significantly, military intelligence and the local people pointed out that roads on the Tibet side are good and enabled smooth flow of traffic. In terms of defence, India has stepped up its troops' presence in the border but the State of the roads makes logistics and supply lines difficult to maintain. The dangerous air routes through the mountains also render it difficult for air traffic to flow in this treacherous mountainous terrain.

Way Forward

Despite these challenges, there is a way forward for better future in the tribal State of Arunachal Pradesh. This road-map could be based on following five pillars.

First, bad governance is not due to the lack of funds. In comparison to other States in India, Arunachal Pradesh's per capita income is favourable. However, for long term prosperity and self reliance; Arunachal Pradesh needs to harness its own resources. For this, following four ways could be done:

(i) Utilise its large forest cover, incidentally the highest in India. Out of 83,743 square km, nearly 68,847 square km (82.21%) of the State is under forest cover. Certain scenic spots, besides Tawang, like Ziro, Mechuka, Tuting, Along, Dibang, etc. should be made tourist friendly and easily reachable.

(ii) Curtail local corruption which is massive by vigilant central oversight. The Arunachal Pradesh Prevention of Corruption (Amendment) Act 1989 is a good mechanism to prevent illegal activities[6].

(iii) The Right to Information (RTI) Act of 2005 is an excellent way to stay informed on development schemes, the allocation of funds, and their implementation process. Arunachal Pradesh has one of the highest numbers of RTI applications. It is also a sound mechanism to curtail corruption.

(iv) Generate resources from local handicrafts and industry.

Second, there has to be a better public relations mechanism between the State institutions and the local population with regard to development of the State. Right now, the approach appears 'top down' and hence there is a ground swell of local resistance to some development projects like the construction of mega dams. If there is better communication on the benefits of various schemes/projects including dams for the State in terms of resource generation and financial self reliance, coupled with sound information on environmental safety checks, the local people will welcome it.

Third, incentivise deputy commissioners who perform best in terms of district development by openly acknowledging their efforts and creating an award system of recognition for the best developed district.

Fourth, Arunachal Pradesh has an immense potential for eco-tourism. Adventure sports, mountaineering, camping and rafting can be promoted to enable tourists to fully enjoy the State's natural beauty. Besides the Tawang monastery, there are several other monasteries which could draw tourists with proper infrastructure development. Serious thought should also be given to remove obstacles to tourism like the PAP as is being done this year on an experimental basis for a period of one year in Manipur, Mizoram and Nagaland.

Fifth, connectivity to Arunachal Pradesh needs to be improved across the international border. The old Ledo Road also called the Stillwell Road during World War II connecting Assam (*via* Margherita and Lekhapani) and Arunachal Pradesh (Lisok and Pangsau Pass on the border) *via* Myanmar to Kunming in China should be looked at as a realistic option for trade connectivity between these two countries and the North-East.

Profile of the Studied Village

The present studied village, Hiya, is situated in the Nyapin sub-division of Kurung Kumey district and is about 18 kilometres away from Nyapin Town, the sub-division headquarters. Hiya, the largest village (in terms of area and population) of Kurung Kumey district is located in the extreme eastern part of the Nyapin sub-division, one of the oldest administrative centre of the State, with whom a status of small town was declared in the same name in 1953. The village has a population of more than 1,000 souls in 308 households as per the recent revised electoral rolls of 2011. This excludes the population below 18 years of age.

In fact, the Hiya village is the blend of two villages, i.e., Hiya-I and Hiya-II and one adjoining sub-village, i.e., Lumtey. Hiya, once a single village was bifurcated into two villages in the year 2001 where the adjoining sub-village Lumtey shift under Hiya-I. The administrations of the village are run under Panchayati Raj Institutions (PRIs) with 2 Anchal Samiti Members (ASMs) and 8 Gram Panchayat Members (GPMs). Besides, the village too has 2 Head Gaon Burahs (HGBs) with few numbers of Gaon Burahs (GBs). The mosaic of these two combinations looks after the administrations of the village.

Table 11.8: Distribution of Households by Type of House

Sl. No.	*Type of House*	*No. of Households*
1.	Traditional Houses	257
2.	Modernised Houses	51
Total		**308**

Source: Block Development Office, Nyapin Sub-Division.

Infrastructure Facilities in Hiya Village

There is a two thirty-seven seater APSTS[7] buses and some private owned Tata Sumos that runs across Hiya village to Ziro and Itanagar respectively. These are only the means of transportation for the people of the village. There is only one IMMARSAT public telephone booth facility in the entire village. However, the mobile phone connection is found in a location 3 kilometres away from the village. The village has post office facility located in the centre of the village near roadside. It also has two fair-price shops under the public distribution system

(PDS) to supply the necessary commodities to the residents at concessional rates. The village has now one government aided newly upgraded secondary school with 14 teachers, one centrally sponsored primary school under Integrated Rural Development Programme (IRDP) and two pre-primary community schools. For higher secondary education they have to go to Nyapin, the sub-division headquarters. One of the reasons for the drop-out of students after the secondary schooling is the difficulty of travelling the distance of 18 kilometres everyday. More often girls are badly affected due to non-availability of higher secondary school in the village. Health sub-centre is there in the village but without a doctor. In many instances of emergency medical cases, they have to go to Community Health Centre (CHC) located at Nyapin to avail the proper medical facilities. One ANM is present in village sub-centre but CHC personnel at Nyapin rarely visit the village for health check-up. Electricity is connected to all households under Rajiv Gandhi Grameen Vidyuthikaran Yojana (RGGVY). Drinking water is available to all the people of Hiya under Rural Water Supply scheme by Public Health Engineering Department (PHED).

Livelihoods of the People in Hiya Village

In Hiya village, the majority of Nyishis are small and marginal farmers where they prefer both shifting and settled cultivations. Traditionally, the indigenous tribal people of this village practice shifting (*jhum*) cultivation, which is locally known as *Tump-Rungho*. The economy of the people is mainly agriculture, and as the major portions of the village are hilly and knotty terrains, they continue to practice this primordial shifting cultivation in hill slopes to fulfil their daily necessities of food and shelter.

Table 11.9: Available Livelihoods in the Village

Sl. No.	*Type of Livelihoods*	*No. of Households*
1.	Agriculture	241
2.	Job	36
3.	Contract	25
4.	Business	06
Total		**308**

Source: AFA[8] Office, Hiya.

From Table 11.9 showing the available livelihoods of the Nyishi people in the village, it may be observed that the households depending on agriculture are more among Nyishis because they have less opportunity to go for other occupation. The livelihoods activities of Nyishis are agricultural cultivation, followed by job, contract, and business (running own shops and hotels). Agriculture is their main source of livelihood followed by job, contract, and business. Mithun, cow, goat, pig, etc., are the major livestock in the village. Individual households raise *puruk* (domestic fowls) in their houses.

Table 11.10: Livestock in the Village

Sl. No.	Particulars of Livestock	No. of Livestock in the Village
1.	Mithuns (Bos Frontalis)	217
2.	Cows	178
3.	Pigs	486
4.	Goats	741
5.	Poultry	3729
Total		**5351**

Source: Fieldwork, Hiya Village.

Table 11.10 on the livestock position in the village shows that the people possess more poultry (3,729) than others in the village. It is followed by goats (741), pigs (486), mithun (217), and cows (178).

Ration

The issue of ration cards indicate the economic status of a household as those who are under below poverty line (BPL), are eligible to get the ration under PDS Scheme in village. It is observed from the PDS records that almost every household have the ration cards. The problem lies in the fact that though majority of the people depend on Public Distribution System (PDS), the dealer has not been supplying the ration items to the people.

Various Development Strategies at Hiya Village

Government of India and State of Arunachal Pradesh is implementing a number of Centrally Sponsored Schemes (CSS) and State sponsored schemes in the areas of rural development, health and family welfare, education, agriculture, women and child development, sanitation, housing, safe drinking water, irrigation, social welfare throughout the Kurung Kumey district, including Hiya village. The main objectives of all these schemes are to generate employment, reduce poverty and economic inequality and improve the quality of life. Besides, some of these schemes aim at creation of basic infrastructure and assets essential for economic development in rural areas. The following categories of development programmes have been implemented in Hiya village for last few years by various departments:

Development Programmes by Department of Planning

District Decentralised Planning Fund

The concept of Decentralised Planning had been introduced in the Kurung Kumey district very recently. The District Planning and Development Boards have full powers to efficiently and cost-effectively implement the district level schemes and also identify the areas and groups of people at the grass root level, which need special attention for equitable socio-economic growth. The District Planning and Development Boards are competent to select executing agencies for executing the works with decentralised funds at their own convenience and for works upto Rs. 20 lakhs, administrative approval will be accorded by these Boards. It is clarified that upper limit is applicable to each individual work and not to the total amount approved under a particular scheme/programme. The works upto Rs. 50 lakhs shall be sent to the Department of Planning for administrative approval. In consonance with the 73rd Amendment to the Constitution, efforts are being made to transfer the fund, function and functionary to the Panchayati Raj Institutions by constituting District Planning Committees.

MLA Local Area Development Scheme (MLALADS)

This scheme enables each Member of Legislative Assembly (MLA) to undertake small developmental works in his/her constituency through the allocated funds of Rs. 2 crore per year. The works recommended under this scheme are conforming to the general pattern of programmes and projects being implemented by the local bodies.

These works are sanctioned and implemented in the same manner as the other works. Wherever require technical and administrative sanctions are given after following the departmental procedures applicable to the local bodies and other government departments. Only that works which can be completed in one or two years and lead to the creation of durable assets should be executed where each individual work should not normally exceed Rs. 70 lakhs. In Hiya village, same funds provided by present MLA are utilising in the construction of school buildings and toilets at village.

MP Local Area Development Scheme (MPLADS)

Under this scheme, funds amounting to Rs. 5 crore per year are placed at the disposal of a Member of Parliament (Rajya Sabha and Lok Sabha) of a Parliamentary Constituency. The scheme is implemented on the guidelines received under the scheme from the Government of India. The funds are released for the works recommended by the concerned M.P. and the works are got executed through the Line Departments/ Implementing Agencies like Panchayati Raj, BDPOs, etc.

Non-Lapsable Central Pool of Resources (NLCPR)

In pursuance to the decision of the Government to earmark at least 10 per cent of Gross Budgetary Support (GBS) of plan funds for the North-Eastern States including Sikkim, the Ministry of Rural Development has been making allocation of plan funds under various rural development programmes. The decision to earmark 10 per cent of Plan Budget was to guarantee availability of funds to the North-Eastern States for implementation of vital pro-poor programmes. Further, it was also made mandatory to transfer the unutilized funds from the earmarked 10 per cent plan budget, each year, to Non-Lapsable Central Pool of Resources (NLCPR) which is administered by the Ministry of Development of North-Eastern Region (DONER).

Development Programmes Department of Relief and Rehabilitation

For Fire Damage: To meet up all relief cases affected by fire accident and other natural calamities. All the relief cases affected by man-made and natural calamities should immediately report with 24 hours for immediate relief. All relief cases will be entertained in district within preview of Deputy Commissioner. All relief cases genuinely scrutinised in cadre under the chairmanship of Circle Officer in the circle and submitted at district. In case of fire accident, report in form of WT message should reach in district within 24 hours for immediate relief and Central Government after discussion may grant Rs. 4,000 to Rs. 6,000 for each household.

For Crop damage: In a year only one time circle level relief committee should hold meeting and scrutinise genuinely.

For Flood Damage: All Flood Damage Report (FDR) will be entertained in district through discretionary power of Deputy Commissioner. It basically depends on financial position.

Development Programmes by District Rural Development Agency

Mahatma Gandhi National Rural Employment Guarantee Act (MGNREGA)

Mahatma Gandhi National Rural Employment Guarantee Act (MGNREGA) is renamed scheme of previously existed National Rural Employment Guarantee Act which was enacted by the Parliament as an Act No. 42 of 2005. The Act provides a guarantee for rural employment to households whose adult members volunteer to do unskilled manual work not less than 100 days of such work in a financial year in accordance with the scheme made under the Act.

Table 11.11: Beneficiaries under MGNREGA Scheme in Hiya village

Sl. No.	*No. of Registered Households*	*Male*	*Female*
1.	250	365	383

Source: DRDA, Koloriang/Kurung Kumey.

Indira Awaas Yojana (IAY)

Indira Awaas Yojana (IAY) is a centrally sponsored scheme funded on cost-sharing basis between the Government of India and State Government in the ratio of 75 : 25. Under this scheme Rs. 25,000 are provided for construction of a House, Kitchen, Smokeless Chulha and Toilet to a family living Below Poverty Line in rural areas in lump sum.

Pradhan Mantri Gram Sadak Yojana (PMGSY)

Rural road connectivity is not only a key component of rural development by promoting access to economic and social services and thereby generating increased agricultural incomes and productive employment opportunities in India, it is also as a result, a key ingredient in ensuring sustainable poverty reduction.

The primary objective of the PMGSY is to provide Connectivity, by way of an All-Weather Road (with necessary culverts and cross-drainage structures, which is operable throughout the year), to the eligible unconnected Habitations in the rural areas, in such a way that all Unconnected Habitations with a population of 1,000 persons and above are covered in three years (2000-2003) and all Unconnected Habitations with a population of 500 persons and above by the end of the Tenth Plan Period (2007). In respect of the Hill States (North-East, Sikkim, Himachal Pradesh, Jammu & Kashmir, Uttaranchal) and the Desert Areas (as identified in the Desert Development Programme) as well as the Tribal (Schedule V) areas, the objective would be to connect Habitations with a population of 250 persons and above.

Swarnajayanti Gram Swarozgar Yojana (SGSY)

The families living Below the Poverty Line belonging to rural areas are assisted under this programme– individually as well as in Groups (Self Help Group). Funds to be received under the scheme are shared by Centre and State Government in the ratio of 75 : 25 ratios. Assistance is provided for income generating activities. After 6 months of the formation of a Self Help Group Rs. 10,000 is provided as Revolving Fund and after completion of second grading (one year) subsidy at 50 per cent subject to a maximum of Rs. 1,25,000 is provided to a Self Help Group.

Sampoorna Grameen Rozgar Yojana (SGRY)

Two erstwhile schemes namely EAS and JRY were reviewed and re-casted as "Sampoorna Grameen Rozgar Yojana" (SGRY) in 2001. This is a centrally sponsored Wage Employment Scheme, Implementation of which will be done through Panchayati Raj Institutions (PRIs). The scheme has the objectives to:

(i) To provide additional wage employment in rural areas,
(ii) To provide food security,
(iii) To create durable community, social and economic infrastructure in rural areas,
(iv) Special emphasis to provide Wage Employment to:
   (a) Women,
   (b) Scheduled castes,
   (c) Scheduled Tribes, and
   (d) Parents of children withdrawn from hazardous occupations.

DDP Watershed Development Projects/Hariyali Scheme

The objective of each watershed development project is to promote the economic development of the village community which is directly or indirectly dependent on the watershed and to encourage restoration of ecological balance in the village. It also includes Development of Agricultural lands, horticulture, grassland, forest land, soil and water conservation measures, creation of water resources, etc. It is a four or five year project and during these period funds amounting to Rs.30 lakhs for one watershed development project are allocated for different components (works, training, community organization, entry point activity, etc.) and an area of 1,250 acres (Approximately 500 hectares) is covered under this project.

Integrated Rural Development Programme (IRDP)

Implementation of the IRDP provided income generating assets and self-employment opportunities to BPL families in the village since the last few years. Most of the families purchased assets belonging to the primary sector. These included pack animals (mules), cows and mithuns and pigs. Some beneficiaries set up shops. There is not much scope for the secondary sector activities in the area. Majority of the beneficiaries was, of course, found happy with their new assets that have increased the flow of income to the family. This activity has created not only self-employment for the beneficiary but also wage employment opportunities for a few others.

Integrated Wastelands Management Programme (IWMP)

Having realised the gravity of the problem of natural resource degradation and the urgency of evolving a strategy for management of the land, various Centrally Sponsored Schemes of Watershed Programme has been taken up in village. With the implementation of Watershed Programmes, the vast *jhum* land which lies barren, are being reclaimed for afforestation, horticulture, cash crop cultivation thereby providing subsistence for the farmers and also various soil and moisture conservation measures are being incorporated in the programme. The programme being community based a sustained community action for operation and maintenance of assets have been created and further development of the potential of the natural resources in the watershed is encouraged. The poorer sections of the society are also benefited by enrolling themselves as part of the Self-Help Group, User group whereby various income generating activities are taken up.

The DRDA, Koloriang with Block Development officers as PIAs under department of Rural Development has been implementing Centrally Sponsored Scheme of Integrated Wastelands Management Programme (IWMP) under which grants-in-aid is provided from the Ministry of Rural Development, Department of Land Resources for development of wastelands on watershed basis directly to DRDAs. This is one of the major watershed programme implemented in the State.

Public Distribution System (PDS) by Department of Civil Supplies

The Government of Arunachal Pradesh introduced the subsidised rise scheme in late 90s to improve the consumption levels of the weaker sections of the society. Since then, a poor household is entitled to 15 kg of rice per person per month at Rs. 3.50 per kg. Besides rice, they are entitled to sugar and kerosene on subsidised rates.

Almost all families have been issued ration cards in the village. It is observed that though the PDS shop is reserved for the people, they are not running the shop. All the people of the village have complained about the increase in the prices of ration items like kerosene, rice, and sugar. They also complained about the quantity of items, which have reduced drastically over a period of time. It is because of the frequent change of norms by the government as well as problem in weighing machine of the dealer/shopkeeper.

Besides the above, the government has been keen to bring development in the areas of education, health, economy, and political participation and so on through Five-Year Plans implemented through various schemes and programmes. Primary and pre-primary schools in the village were established to bring the educational development among the villagers. Similarly, health sub-centre was also established and health functionaries were to periodically visit the village to extend health facilities. Thus, the approach of government towards development has been comprehensive and holistic in nature.

People's Participation in Developmental Process

People's participation in local level planning means participative development. In its ideal from, local planning implies entrusting to the local people and their institutions all duties and responsibilities of local planning and development, with the government reserving to itself only the functions of guidance, supervision and higher planning. In fact, the local level planning process is one of working with the local people, most directly affected by facilities and programmes, so that the plan: is more responsive to local needs; reflects more accurately local perceptions; produces a sense of local ownership and responsibility; builds on and reinforces the fabric of the community and its internal structure; and supports the evolution of a continuing on-going structure of local administration by creating a core of people.

These people having worked to make a project happen, and having seen ideals become a part of development, would carry their involvement into administration, maintenance and planning for the future. Elected members of the Panchayat Samitis and Gram Panchayats, are now directly involved in implementation of the development programmes. Gram Panchayats also prepare a shelf of need-based projects which, after sanction at the Block and District level, are executed by them with the funds made available to them. In short, people are actively participating in the development process directly and through their elected representatives.

Analysis of the Development Schemes

In the following an attempt is made to analyse the implementation of the above schemes and their outcomes in Hiya village. It will be seen that there have been several problems in implementing the schemes. Thus, the objectives of the schemes have partially met.

In spite of the above-mentioned efforts of Government, people's position in Hiya village has not changed significantly. One dominant reason for such static conditions in spite of efforts made to effect change is obviously as in any other parts of district and State due to the apathy of the government functionaries at different levels. Local conditions such as poor infrastructural development, failure of monsoons, interests of the communities are also to some extent responsible for the inertia.

It is contended that even the village Panchayat does not perform its function objectively. It is believed that the Panchayat is favouring one section of the people and delivers all its services to the same section of the people and deliberately neglecting the other. This is due to the party politics prevalent in the villages. Some villagers opposed to the Panchayat's stand but it was of no use. Thus, the party politics have a direct bearing on the development efforts, either by Panchayat or government machinery.

The levels of interaction of Nyishi villagers with both the government officials can be seen in terms of their involvement in the party politics and factions. Ruling party is always ahead in the factional politics in the villages and implement agencies have to take sides with the ruling party or faction. Almost all current incumbencies in Panchayat in the village were elected on Indian National Congress (INC) party ticket.

The above-described perspective revealed the nature of development processes in the studied village. Even though there is no significant socio-economic change of the people in the village, it cannot be denied some qualitative change in the lives of the Nyishi people. In terms of quantity that is the number of households who cross the below poverty line due to the efforts of the government, there is initiation of change among Nyishis of the Hiya village which may take some time to yield perceptible results. Some micro level changes, which are qualitative in nature likely to produce results, have been noticed in the study.

The panchayat seem to have successfully brought the awareness and institutional means of savings among the Nyishis. Though the government provided the conceptual framework, it is Panchayat, which completed the process for a success. The resources available in the village primarily are land, water and water resources with few human resources. The Panchayat recognised the availability of youth who could be used for capacity building purposes. This programme is successful as the youth though a small number were able to make their own living better than those who depended entirely on the farming.

It is no doubt that the government approach has been integrative but there is poor coordination of various departments for various reasons. The holistic development has been conceived but due to weakness in delivery systems the end results is pathetically obtusely stagnant. But, consistent efforts made, with all inherent problems focusing on economic change while consciously taking care of supportive cultural capital, has yielded some positive results. As the village is part of the wider political system it cannot escape from the good and bad aspects of the system. Some negative effects are visible hampering the development of the local population, which requires solution at different levels beyond the village.

Conclusion

There is only one active player, i.e., government, in bringing planned social and economic change leading to development of Nyishi tribal people, inhabiting in Hiya village. The Nyishis have been poverty stricken, largely remained farmers needed the intervention of government for being rescued from their deplorable conditions.

In order to understand the impact of the interventions of the above agencies to bring socio-economic development, the study has taken people's perspective of analysing development process. The development programmes and schemes at macro level are broad encompassing various issues, integrative and ideal efforts to stamp out poverty by creating infrastructures and offering support to individuals and individual households. At the micro level, the programmes are beset with problems relating to coordination of various elements that necessarily intervene and intersect the areas of operation. These include human elements—discharging the duties of the functionaries, location of the institutions, power politics and natural local conditions. There is a complex relationship of these elements. For success of any programme, proper configuration, manipulation and exploitation of these elements for advantage become necessary. Whoever, either an individual or agency, is able to handle judiciously and overcome these problems will be a successful player. Economic focus ensuring support of cultural elements seems to yield desired result, as can be finely drawn from present study.

The fundamental challenges facing the government and the civil society in Arunachal Pradesh today include the tasks of secularising governance and democratising development (Upadhyay *et al.*, 2004). The goals of building capabilities, expanding choices and safeguarding freedoms cannot be achieved unless the development process is made more inclusive and participatory in nature. The often talked about but least implemented, bottom-up approach to development is likely to suit the aspirations of the people in this State with so much cultural and ecological diversity. However, there is an urgent need to develop a transparent and non-discriminatory framework for governance, which in turn requires effective monitoring mechanisms and accountability at all levels. While there is always a scope for enhancing public investment in the development of physical and social infrastructure, mere announcement of financial packages, without an institutional strategy that ensures proper utilization of the fund will only reinforce the existing cycles of corruption, inefficiency and perpetual dependence. The role of civil society in this transformation can hardly be overstressed.

The question of identity and cultural distinctiveness of the indigenous people in the State is an issue that needs urgent attention and its relationship with the political economy of development has to be examined at greater detail. In Arunachal Pradesh, the key challenge in developing a non-discriminatory framework for governance, without compromising with the basic collective rights of the indigenous population, is to be able to address particularistic demand within the framework of universal norms. This, by all counts, is seems to be a tough challenge before the decision-makers/policy makers.

REFERENCES

Adelman, I. (1961). *Theories of Economic Growth and Development*. UK: Stanford University Press.

Bezbaruah, M.P. and M.K. Dutta. (2001). Economic Transition in the North-Eastern Region Before and After Liberalisation. *Assam Economic Journal,* Vol. 14, pp. 17-31.

Bhagabati, A.C. (1990). Anthropological Perspectives on Tribal Development in the Assam Hills: A Preliminary Statement. In B.N. Bordoloi (ed.), *Constraints of Tribal Development in North East India*. Guwahati: Tribal Research Institute, Assam.

Bhattacharjee, P.R. (2002). The Problem of Ensuring Financial Efficiency in the North Eastern States. *Dialogue,* Vol. 4, No. 1, pp. 59-74.

Chambers, R. and R. Conway. (1992). *Sustainable Rural Livelihoods: Practical Concepts for the 21st Century.* Brighton, UK: Institute of Development Studies (IDS), Discussion Paper No. 296.

Chaudhuri, I.N. (2009). *Dilemma of the Multilingual Society and the Culture of Development: The Indian Scene.* In M.K. Sanyal and A. Ghosh (eds.), *Culture, Society and Development in India.* New Delhi: Orient Blackswan.

Choudhury, S.D. (1981/2008). *Gazetteer of India: Arunachal Pradesh (Subansiri District).* Itanagar: Government of Arunachal Pradesh/ Himalayan Publishers.

Clark, C. (1951). *Conditions of Economic Progress* (2nd Edition). London: Macmillan

Deogaonkar, S.G. (1994). *Tribal Administration and Development: With Ethnographic Profiles of Selected Tribes.* New Delhi: Concept Publishing Company.

Department of Economics & Statistics (2011). *Statistical Abstracts of Arunachal Pradesh: 2010-11.* (2011). Itanagar: Government of Arunachal Pradesh.

Desai, S.B.; A. Dubey; B.L. Joshi; M. Sen; A. Sharif & R. Vanneman. (2010). *Human Development in India: Challenges for a Society in Transition.* New Delhi: Oxford University Press.

Drewnowski, Jan. (1966). *Social and Economic Factors in Development.* Geneva: UNRISD, Paper No.3.

George, J. and S.S. Sree Kumar. (1994). *Tribal Development: Legislation and Enforcement.* New Delhi: Commonwealth Publishers.

Gide, C. and C. Rist. (1959). *History of Economic Doctrines.* Translated by Richards. UK: George G. Harrap & Co.

Government of Arunachal Pradesh (2006). *Arunachal Pradesh Human Development Report-2005.* Itanagar: Department of Planning, Government of Arunachal Pradesh.

Government of India (2010). *Tourism Development Plan: Arunachal Pradesh.* New Delhi: Heritage Tourism Division, Indian National Trust for Ministry of Tourism Art and Cultural Heritage (INTACH).

Gupta, D. (2004). *Meeting "Felt Aspirations": Globalization and Equity from an Anthropological Perspective.* Retrieved on August 2, 2009 from www.anthrosource.net/doi/abs/10.1525/Gupta.2004/pdf.

Jain, H. (2004). *Scheduled Tribes—Changes in Socio-Economic Conditions.* New Delhi: Classical Publishing Company.

Jayal, N.G. (1997). The Governance Agenda: Making Democratic Development Dispensable. *Economic and Political Weekly,* Vol. 32, No. 8, pp. 407-412.

Jha, S.D. (1988). *Socio-Economic and Demographic Dimensions of Arunachal Pradesh.* New Delhi: Omsons Publications.

Luthra, P.N. (1971). *Constitutional and Administrative Growth of North East Frontier Agency* (NEFA). Shillong: Regency Publisher.

Madhab, J. (1999). Northeast: Crisis of Identity, Security and Underdevelopment. *Economic and Political Weekly,* Vol. 34, No. 6, pp. 320-322.

Maheshwari, S.R. (1985). *Rural Development in India: A Public Policy Approach.* New Delhi: Sage Publications.

Meena, R.S. (2006). *Tribal Development Programmes: A Critical Appraisal.* Jaipur: Ritu Publications.

Mishra, D.K. and V. Upadhyay (2004). Governance and Development in Arunachal Pradesh: The Emerging Challenges. *Dialogue*, Vol. 5, No. 4 (April-June, 2004).

Planning Commission (2002). *The National Human Development Report 2001.* New Delhi: Planning Commission.

Ramaiah, P. (1988). *Issues in Tribal Development.* Allahabad: Chugh Publications.

Roy Burman, B.K. (1989). Problems and Prospects of Tribal Development in North-East India. *Economic and Political Weekly*, Vol. 24, No. 13 (Apr. 1, 1989), pp. 693-697.

Roy, N.C. (1996). Growth and Structural Changes in the Economy of Arunachal Pradesh. *Arunachal University Research Journal,* Vol.1, No. 1.

Sachdeva, G. (2000). *Economy of the North-East: Policy, Present Conditions and Future Possibilities.* Centre for Policy Research, New Delhi: Konark Publishers.

Seers, D. (1969). The Meaning of Development. *International Development Review,* Vol. 11(4), pp. 3-4.

Seers, D. (1972). *What are we Trying to Measure.*UK: IDS.

Somasekhar, K. (2008). *Developmental Programmes and Social Change among the Tribals.* New Delhi: Serials Publications.

Talukdar, A.C. (1990). *Constraints of Rural Development in Tribal Areas: The Arunachal Experience.* In B.N. Bordoloi (ed.), *Constraints of Tribal Development in North East India.* Guwahati: Tribal Research Institute, Assam.

UNDP. (1990). *Human Development Report.* New York: Oxford University Press.

Upadhyay, V. (2002). Violence and Development. *Social Change and Development*, Vol. 1, No.1, pp. 36-43.

Weidner, E.W. (1970) *The Elements of Development Administration.* In Weidner, E.W. (ed.), *Development Administration in Asia.* Durham, N.C.: Duke University Press.

NOTES

1. Denis Goulet was a pioneer of human development theory and the main founder of work on "development ethics" as a self-conscious field that, by his definition, treats the ethical and value questions posed by development theory, planning, and practice.
2. See "DoNER Minister Review NLCPR/NEC Schemes", *The Morung Express*, April 27, 2011 at *http:// www.morungexpress.com/regional/ 65300.html* (Accessed on June 5, 2012).
3. "Tourism Development of Arunachal Pradesh" Ministry of Tourism, Government of India, February 2010 at *http://tourism.gov.in/survey/ ArunachalPradesh-Feb-10.pdf* (Accessed on June 5, 2012).
4. See "Arunachal State Budget presented", Arunachal News, March 19, 2011 at *http://arunachalnews.com/arunachal-state-budget-presented.html* (Accessed on June 5, 2012).
5. *For more on hydro-electric projects in Arunachal Pradesh, see "Hydro-Electric Power Policy Arunachal Pradesh" at* http://www.arunachal hydro.org.in/pdf/State%20Mega%20Hydro.pdf *(Accessed on June 5, 2012).*
6. "The Arunachal Pradesh Prevention of Corruption (Amendment) Act, 1989" at *http://www.lawsofindia.org/statelaw/1482/TheArunachal PradeshPreventionofCorruptionAmendmentAct1989.html* (Accessed June 5, 2012).
7. *It stands for* Arunachal Pradesh State Transport Services.
8. It stands for *Agricultural Field Assistant.*

12

Population Growth, Urbanisation and Sustainable Development: A Manipur Perspective

Elangbam Nixon Singh

Introduction

Urbanisation, simply defined, is the shift from a rural to an urban society, and involves an increase in the number of people in urban areas during a particular year. Urbanisation stands for growth of towns and non-agricultural activities. Truly speaking, it is associated with the growth of industrialization, trade and commerce. Urbanisation is the outcome of social, economic and political developments that lead to urban concentration and growth of large cities, changes in land use and transformation from rural to metropolitan pattern of organization and governance.

Urbanisation is increasing in both the developed and developing countries. However, rapid urbanisation, particularly the growth of large cities, and the associated problems of unemployment, poverty, inadequate health, poor sanitation, urban slums and environmental degradation pose a formidable challenge in many developing countries. Available statistics show that more than half of the world's 6.6 billion people live in urban areas, crowded into 3 per cent of the earth's land area (Angotti, 1993; UNFPA, 1993). The proportion of the world's population living in urban areas, which was less than

5 per cent in 1800 increased to 47 per cent in 2000 and is expected to reach 65 per cent in 2030 (United Nations, 1990; 1991). However, more than 90 per cent of future population growth will be concentrated in cities in developing countries and a large percentage of this population will be poor. In Africa and Asia where urbanisation is still considerably lower (40%), both are expected to be 54 per cent urban by 2025 (UN, 2002). Sustainable development requires economic efficiency, social equity, human security and ecological stability and recovery. Substantial transfer of knowledge, technology and financial resources are needed from advanced to less advanced regions to promote sustainable development at a global scale. Several UN report like the 'Global Environment Outlook' (UNEP, 1999), 'Current Trends' (CSD) and 'Bending the Curve' of UNEP (Raskin *et al.*, 1998) had analysed various dimensions of sustainable development. The picture arising from these reports is that the policy emphasis at the global level should be on:

(a) Population, migration and the distribution of health and wealth (social domain);
(b) Biodiversity and nature protection as a proxy for the ecological carrying capacity of the biosphere (ecological domain);
(c) The eco-efficiency of the use of natural resources, in particular, water, land and energy (economic domain); and
(d) The use of effective and efficient policy instruments (institutional domain).

Although urbanisation is the driving force for modernisation, economic growth and development, there is increasing concern about the effects of expanding cities, principally on human health, livelihoods and the environment. The implications of rapid urbanisation and demographic trends for employment, food security, water supply, shelter and sanitation, especially the disposal of wastes (solid and liquid) that the cities produce are staggering (UNCED, 1992). The question that arises is whether the current trend in urban growth is sustainable considering the accompanying urban challenges such as unemployment, slum development, poverty and environmental degradation, especially in the developing countries.

Manipur and its Geographical Location

Manipur, one of the eight sisters of North-Eastern Region of India is situated in the strategic most borders bounded on the north by Nagaland, on the east by Myanmar, on the south by Mizoram and on the west by Assam. The State is located in a rectangular shape between 23°50′ N and 25°42′ N latitudes and 92°58′ E and 94°45′ E longitudes. The geographical area of the State is 22,327 sq. km. Basically, it can be divided as hill and valley. The hill region covers nine-tenth of the area.

Here, it is worthwhile to mention that information regarding the total land utilization of the entire State cannot be obtained since land records are available only for the valley and a small part of the hill areas. Due to the rapid growth population in the State, acute pressure is on the valley area where all the people belonging to different ethnic groups use to settle and concentrate with high density, thereby leading to urbanisation. On the hand, the region accounting for 90.00 per cent of the total land area of the State is only meant for the scheduled tribes and is sparsely populated.

The valley has reached a stage of population explosion. The density in the valley consisting of four districts has been 631 persons per sq. km as against 44 persons per sq. km. in the hill region (as per 2001 Census). It is of the general opinion that Manipur Land Revenue and Land Reforms Act 1960 should be extended to the whole of the State for proper utilization of land areas of the State (Singh, Nixon and Prasain, G.P., 2010). At the same time, the population of the State is composed of different religious communities like Hindu, Muslim, Christian, Sikh, Jain, Buddhist, etc. Though the majority Hindu constitutes 60.00 per cent of the population, 34.00 per cent are Christians. The third majority community is the Muslims (locally known as Meitei Pangals). It is also very interesting to know that a sizeable number of persons who have faith in the tradition Meitei religion better known as "Sanamahi" are recorded recently in the category of "other religions".

Economically, Manipur is a backward State. It can be characterized by the absence of any medium and large scale industries and predominance of backward agricultural practice and allied activities accompanied by large scale shifting cultivation especially in the hill regions.

Literature Review

Dr. Alejandro Canales (1999), in his paper, examines the dramatic population increase occurring along the border, the ways it impacts community health and the environment, and the development model driving much of that growth on Mexico's northern border cities that as of 1990 boasted over 100,000 inhabitants each—Tijuana, Mexicali, Nogales, Ciudad Juárez, Nuevo Laredo, Reynosa, and Matamoros. Their combined population of 2.8 million people accounts for nearly 7 per cent of Mexicans residing in medium-sized cities and metropolitan areas. (In this article, cities with more than 100,000 people are defined as medium-sized, while Mexico City, Guadalajara, Monterrey, and Puebla are defined as metropolises due to the influence they exert on politics and the economy at the national level.) Although the population of northern Mexico's cities has grown substantially throughout the course of the entire century, it has increased at an even faster rate over the last decade. This incessant population growth has generated a number of serious urban problems, including a lack of drinking water, inadequate sewage services, substandard housing, insufficient garbage disposal, and air and water pollution, as well as diverse negative environmental impacts outside city boundaries.

Around the world, especially in Africa and Asia, cities are expanding rapidly. For the majority of urban dwellers, especially the poor, finding potable water supply, affordable shelter, accessible and secure urban land for agriculture to ensure food security, securing gainful employment and improvement in health facilities would continue to remain a priority (Kwasi Nsiah-Gyabaah, 2002).Since restrictive urban growth policies, especially population distribution designed to reduce the rate of rural-urban migration appear to have had limited success in many developing countries, policies must be directed at transforming the rural economy in order to slow the rate of urban sprawl. Comprehensive land-use planning and revision of planning standards and administrative procedures would go a long way to reduce many of the problems that face urban populations in the developing areas, especially Africa. In less developed countries many cities are surrounded by informal settlements of low income groups, with poor access to clean water and urban facilities. Spatial and infrastructure planning are often lacking or lagging behind, which could harm the attractiveness of cities for economic activities, despite the large potential (labour and consumer) market. The World

Development Report 1999-2000 (World Bank, 1999) states that urbanisation and economic growth go hand in hand in all parts of the world, with the exception of Africa. According to the World Bank, the African pattern of urbanisation with growth is caused by the subsidies on food and trade that favoured urban consumers over rural producers. Poverty will thus remain to be the main problem of cities in Africa.

Countries differ not only in the level of urbanisation but also in the size of their urban populations (UNPD, 2001). As the distribution of countries and areas according to urban population size shows, the variation in urban population size has been increasing. Both the unprecedented population growth that many countries experienced during the second half of the twentieth century and their rising levels of urbanisation have resulted in increasing numbers of people living in urban areas.

Demands for higher standards of living in urban areas have made it difficult for traditional leadership structures to respond in ways perceived as adequate by town dwellers. A large and increasing proportion of the urban young has never visited their traditional village communities and have not strong links with them. Thus traditional safety nets developed by Pacific island societies over hundreds of years are becoming increasingly strained as expectations are modified by development and many communities grapple with the transition from subsistence to cash-based livelihoods in growth-oriented economies. As a result, the number of people living below the poverty line in urban areas is increasing (UNPD, 1996).

Urbanisation, or the shift of population from rural to urban environments, is typically a transitory process, albeit one that is socially and culturally traumatic. As a country develops it moves from labour intensive agricultural production to labour being increasingly employed in industry and services. The latter are not land-intensive and are located in cities because of agglomeration economies. Thus urbanisation moves populations from traditional rural environments with informal political and economic institutions to the relative anonymity and more formal institutions of urban settings. That in itself requires institutional development within a country. It spatially separates families, particularly inter-generationally as the young migrate to cities and the old stay behind (J. Vernon Henderson, 2004).

Amitabh Kundu's (2006) paper argued that the process of economic liberalization and associated structural reform would accelerate Rural-Urban (RU) migration and boost the pace of urbanisation. Linking of India with global economy would lead to massive inflow of foreign capital as also rise in indigenous investment resulting in an increase in employment opportunities within or around the existing urban centres.

Objectives of the Chapter

The present study is purely based on various secondary data/reports published by the State government and the census data, being prepared by Census of India, from time to time. In fact, the First Census count in India was found to be partially conducted in 1872 (Mahendra K. Premi, 1991) by covering only a limited area though it was considered to be the foundation of the history of population census in the country. Then, the Second Census was conducted in 1881 covering wider areas and since then regular and periodical population census of the country is found to have been conducting for every ten years. It can be pointed out that the population of Manipur was recorded with only 2,21,070 persons in 1881. Then in the year 1891, the princely state was annexed by the British. Later, the population of Manipur was recorded as 2,84,465 persons with an addition of only 63,395 persons in twenty years (1881-1901). But one of the important features of the trend of population growth is that during the period 1901 to 1951, the population of Manipur reached 5,77,635 persons with a net addition of 2,93,170 persons during the period of 50 years. In other words, it took 50 years to complete a little more than double of the population of 1901.

In this chapter, an attempt has been made to analyze the growth of population in Manipur during 1951 and 2011, keeping urbanisation in view for sustainable development of the State.

Growth of Population in the North-Eastern States

First of all, the growth of population in the entire North-Eastern States of India may be observed to make a comparative study. Table 12.1 shows that Arunachal Pradesh has more or less identical population growth, even though the first Census was started in 1961 in this State. In Assam, the decadal growth rate of population was maximum in 1971-81 (36.05%) and minimum in 1981-91 (12.65%). The highest growth rate can be found in Tripura (78.71%) in 1951-61 which may be due to influx of migrants from Bangladesh (erstwhile East Pakistan).

In Nagaland, the high rate of growth rate was observed since 1971 Census (39.88) and it attained its peak (64.53%) in the just concluded 2001 Census. But surprisingly, in 2011 Census, it had a negative growth rate of (–) 0.47. On the other hand, Assam is the highest populated State (3,11,69,272 persons) and Sikkim, the least one (6,07,688 persons) as per 2011 Census report.

Table 12.1: Decadal Growth of Population of North-Eastern States of India

States	*1951*	*1961*	*1971*	*1981*	*1991*	*2001*	*2011*
Arunachal	–	336558	467511	631839	864558	1097968	1382611
Pradesh		–	(38.91)	(35.15)	(36.87)	(27.00)	(25.92)
Assam	8028856	10837329	14625152	18041248	22414322	26655528	31169272
	(19.93)	(34.98)	(34.95)	(36.05)	(12.65)	(18.92)	(16.93)
Meghalaya	605674	769380	1011699	1335819	1774778	2318822	2964007
	(8.97)	(27.03)	(31.50)	(32.50)	(32.86)	(30.65)	(27.82)
Mizoram	196202	266063	332390	493757	689756	888573	1091014
	(28.42)	(35.61)	(24.93)	(48.55)	(39.68)	(28.82)	(22.78)
Nagaland	212975	369200	516449	774930	1209546	1990036	1980602
	(8.60)	(14.07)	(39.88)	(50.05)	(56.13)	(64.53)	(-0.47)
Tripura	639029	1142005	1556342	2053058	2757205	3100203	3671032
	(24.56)	(78.71)	(36.28)	(31.92)	(34.29)	(16.03)	(14.75)
Sikkim	137725	162189	209843	316385	406457	540851	607688
	N.A.	(17.76)	(29.38)	(50.77)	(28.47)	(33.06)	(12.36)
Manipur	577635	780037	1072753	1420953	1837149	2293896	2,721,756
	(12.80)	(35.04)	(37.53)	(32.46)	(29.29)	(24.86)	(18.65)

Note: Figures in parentheses indicates decadal growth in percentage.
Sources:

(i) Provisional Population Totals of 2011, Office of the Registrar General & Census Commissioner, India.
(ii) Statistical Abstract of Manipur, 2005.
(iii) Economic Review of Manipur (1994-95), Directorate of Economics and Statistics, Manipur, p. 9.

Trend of Population Growth in Manipur

Population-wise, Manipur is a small State as compared to other States of India like U.P. or Bihar. But it can be observed from Table 12.2 that the State has been experiencing a very high population growth over the years since 1951. The population of the State was 5,77,635 persons in 1951 and it reached 7,80,037 persons in 1961 (35.04 per cent growth rate). At the national level, it was only 21.51 per cent.

Table 12.2: Trend of Population Growth in Manipur

Year	*Population*	*Net increase of population*	*Decadal Growth in percentage*		
			Manipur	*North-East*	*All India*
1951	5,77,635				
1961	7,80,037	2,02,402	35.04	41.01	21.51
1971	10,72,753	2,92,716	37.53	34.98	24.80
1981	14,20,953	3,48,200	32.46	26.52	25.00
1991	18,37,149	4,16,196	29.29	27.61	23.85
2001	22,93,896	4,56,747	24.86	22.00	21.54
2011	27,21,756*	427860	18.65	15.01	17.64

Sources: 1. Provisional Population Totals of 2011, Office of the Registrar General & Census Commissioner, India.
2. Statistical Abstract of Manipur, 2005.

* Includes estimated population of Mao-Maram, Paomata & Purul sub-divisions of Senapati district.

The highest growth rate was in 1971 when it was recorded as high as 37.53 per cent against the all India rate of 24.80 per cent. It is also interesting to note that the State also exceeds the overall North-East rate except in the1961 Census. But the decadal growth rate started declining from 1981 Census (32.46%) against 25.00 per cent of all India rate. In the recent Census of 2011, the growth rate was observed at only 18.65 per cent against 17.64 per cent of all India rate.

This analysis shows that the trend of population growth has been not only continuously high in Manipur over the years but also significantly much higher than the national level and North-East level during the last 6 (six) decades under study.

District-wise Growth of Population

The uneven growth of population in the districts of Manipur is also an interesting area of discussion. Table 12.3 observes that Senapati district had, in all the censuses, all time high rates of growth which are above the State averages. There might be many reasons, which can only be known by the concerned respondents. Another scene can be seen in the district of Tamenglong in 1951-61 census. The growth rate was negative (–3.10%) during that period. However, it increased to 23.16 per cent, 38.50 per cent, 38.51 per cent and 29.23 per cent in 1971, 9181, 1991 and 2001 censuses respectively. In Chandel district, the maximum growth rate of 66.62 per cent was recorded in

2001 census which had arisen so many questions.The Churachandpur district also had high rates of growth since 1951 and it never came down below the State averages. It had the highest growth rate in 1961-1971 period with 57.19 per cent. The rate of growth in the hill districts of Senapati, Churachandpur, Tamenglong and Chandel could be because of large scale influx of immigrants into these districts (Dun, E.W., 1975).

On the hand, the growth rates in the valley districts were more or less consistent around the State averages except in Imphal and Bishnupur districts in the 1951-61 period. Imphal and Thoubal stood first and second in the growth rate of population with 75.72 per cent and 67.87 per cent respectively. Later on, in all other censuses, the valley districts had never exceeded the growth rates of the State averages for all the time.

Table 12.3: District-wise Decadal Growth Rate of Population (1951-2001)

District	*1951-61*	*1961-71*	*1971-81*	*1981-1991*	*1991-2001*	*2001-2011*
Senapati	66.80	44.61	49.19	34.09	36.09	25.16
Tamenglong	(-)3.10	23.16	38.50	38.51	29.23	25.69
Churachandpur	46.20	57.19	37.37	30.72	29.36	19.03
Chandel	15.09	39.90	45.76	25.81	66.62	21.72
Imphal@	75.72	34.72	27.51	26.83	@	*
Thoubal	20.95	36.19	28.01	27.89	23.87	15.48
Bishnupur	67.87	37.35	30.33	28.07	15.27	15.36
Ukhrul	14.35	28.07	33.29	31.74	28.83	30.07
Total	**35.04**	**37.53**	**32.46**	**29.29**	**24.86**	**18.65****

Note: Imphal District was bifurcated into Imphal East and Imphal West Districts in 1997.

@ Imphal East – 19.49   * Imphal East – 14.63

Imphal West – 16.70   Imphal West– 15.82

** Includes estimated population of Mao-Maram, Paomata and Purul sub-divisions of Senapati district.

Source: Statistical Abstract of Manipur, 2001 Directorate of Economics and Statistics, Manipur.

1.8 Growth of Population by Religions

Another aspect of population study in Manipur is the growth of population as characterized by different ethnic and religious communities. In fact, the State is considered as the home land of the Meiteis, who are by and large Hindus, and different tribes and

Manipuri Muslims known as Meitei Pangals who had migrated a few centuries ago. Nowadays, there are also the Bengalis, Biharis, Nepalese, Jains, Sikhs, etc., who had migrated and settled in the State for livelihood. The majority of the Meiteis are valley dwellers, though a few of them are also found in certain areas of the hill region. The Hindus accounted highest (61.68%) in 1961 and lowest in 2001 (46.01%). It may be due to the fact the Meiteis who had faith in "Sanamahi" are recorded in the "other" religion from 2001 Census.

Next to Hindus, the Christians are the second largest communities accounting for 11.84 per cent in 1951. But it increased to 34.11 per cent and 34.04 per cent in 1991 and 2001 respectively. They are mainly hill dwellers but nowadays a major chunk of them is also found in the valley.

Table 12.4: Population by Religions in Manipur (1951-2001)

Religions	*1951*	*1961*	*1971*	*1981*	*1991*	*2001*
Hindu	3,47,325 (61.13)	4,81,112 (61.68)	6,32,597 (58.97)	8,53,180 (60.04)	10,59,470 (57.67)	9,96,894 (46.01)
Christian	98,394 (11.84)	1,52,043 (19.49)	2,79,243 (26.68)	4,21,702 (29.68)	6,26,669 (34.11)	7,37,578 (34.04)
Muslim	37,197 (6.44)	48,588 (6.23)	70,969 (6.62)	99,327 (6.99)	1,33,535 (7.27)	1,90,936 (8.81)
Sikh	*50	523 (0.07)	1,028) (0.10)	992 (0.07)	1,301 (0.07)	1653 (0.07)
Buddhist	*33	325 (0.04)	495 (0.05)	473 (0.03)	711 (0.07)	1926 (0.08)
Jain	150 (0.03)	778 (0.10)	1408 (0.13)	975 (0.07)	1337 (0.07)	1461 (0.06)
Others	1,24,486 (21.55)	–	83,167 (7.75)	35,490 (2.50)	14,066 (0.77)	2,35,280 (10.86)
Religion not stated	–	96,668 (12.39)	3,846 (0.36)	8,814 (0.62)	–	1057 (0.05)
Total	**5,77,635 (100.00)**	**7,80,037 (100.00)**	**10,72,753 (100.00)**	**14,20,953 (100.00)**	**18,37,149 (100.00)**	**21,66,788** (100.00)**

Note: Figures in parentheses indicate percentage to total.

* Negligible.

** Figures of 2001 Census excludes Mao-Maram, Paomata and Purul Sub-Division of Senapati District.

Sou ces: (i) Economic Survey of Manipur, 1999-2000, and
(ii) Statistical Abstract of Manipur, 2005.

The third important religious community is the Muslims. Their numbers are less than 10.00 per cent of the total population of the State (Table 12.4). Like Meiteis, Muslims of Manipur (Meitei Pangals) are valley dwellers. All the Muslims of Manipur are not Meitei Pangals because Meitei Pangals are those who are born and brought up in Manipur. Even among the Meiteis, some of them are Scheduled Caste (2.00%), *viz.*, Lois, Yaithibi, Dhobi, etc. It is worthwhile to mention here that the population of Manipur is divided into eight religious groups, *viz.* Hindus, Christians, Muslims, Sikhs, Jains, Buddhists, others, and religions not stated as above.

Urbanisation in Manipur—An Analysis

Table 12.5 analyse the rate of urbanisation in Manipur as compared to other States of North-Eastern Region of India. Manipur ranks second to Mizoram in the process of urbanisation according to both 2001 and 2011 censuses.

Table 12.5: Urbanisation in North-Eastern States of India

State/India	*1971*	*1981*	*1991*	*2001*	*2011*	*Urban Population*	
						2001	*2011*
Arunachal Pradesh	3.70	6.56	12.80	20.75	22.67	227881	313,446
Assam	8.82	9.88	11.10	12.90	14.08	3439240	4,388,756
Manipur	13.19	26.42	27.52	25.11	30.21	575968	822,132*
Meghalaya	14.55	18.07	18.60	19.58	20.06	454111	595,036
Mizoram	11.36	24.67	46.10	49.63	51.51	441006	561,997
Nagaland	9.95	15.52	17.21	17.23	28.97	352821	573,741
Sikkim	9.37	16.15	9.10	11.07	24.97	59870	151,726
Tripura	10.43	10.99	15.30	17.06	26.18	545750	960,981
India	19.91	23.34	25.70	27.81	31.16	286124165	377,105,760

Source: (a) Selected Socio-Economic Statistics, India, 2006.
(b) Provisional Population Totals of 2011, Office of the Registrar General & Census Commissioner, India.

* Includes estimated population of Mao-Maram, Paomata and Purul sub-divisions of Senapati district.

Natural population increase (high births than death) and migration are significant factors in the growth of cities of India. The case is also the same with Manipur. The natural increase is fuelled by improved medical care, better sanitation and improved food supplies, which reduce death rates and cause populations to grow. In many developing countries, it is rural poverty that drives people from the rural areas into the city in search of employment, food, shelter and education. Data show the poor trend to have larger families[1]. In Manipur, most people move into the urban areas because they are 'pushed' out by factors such as poverty, environmental degradation, religious strife, political persecution, food insecurity and lack of basic infrastructure and services in the rural areas or because they are 'pulled' into the urban areas by the advantages and opportunities of the city including education, electricity, water, etc. Even though in many urban areas offer few jobs for the youth and they are there often attracted by the amenities of urban life.

According to 2011 Census, a total of 822,132 populations were found in the urban area constituting 33 towns (28 statutory and 5 census towns) in Manipur (Table 12.6).

Table 12.6: Urbanisation in Manipur (1901-2011)

Year	*No. of Towns*	*Urban Population*	*% of Total Population*	
			Urban	*Rural*
1901	1	72234	25.39	74.61
1911	1	74650	21.56	78.44
1921	1	80003	20.83	79.17
1931	1	85804	19.26	80.74
1941	1	99716	19.47	80.53
1951	1	2862	0.50	99.50
1961	1	67717	8.68	91.32
1971	8	141492	13.19	86.81
1981	32	375460	26.42	73.58
1991	31	505645	27.52	72.48
2001	33	575968	25.11	74.89
2011	33	822,132*	30.21	69.79

Source: (a) Economic Survey, Manipur 2007-08 p.12.
(b) Provisional Population Totals of 2011, Office of the Registrar General & Census Commissioner, India.

* Includes estimated population of Mao-Maram, Paomata and Purul sub-divisions of Senapati district.

It appears from Table 12.6 that the process of urbanisation has been very slow in the State which perhaps indicates a slower occupational shift. The number of towns and cities has grown at a low rate from 1 (one) town in 1951 to 33 in 2011 during the last 50 years period. However, at a closer analysis, the urban population has increased from about 72.234 thousands to 822.132 thousands over the said period.

Processes of Urbanisation

One significant feature of the urbanisation process in Manipur is that unlike other cities of India, much of the growth is taking place in the absence of significant industrial expansion. Urbanisation also finds expression principally in outward expansion of the built-up area and conversion of prime agricultural lands into residential and industrial uses. An alternative to the present expansion of the urban population across the State in order to save prime land for agriculture is to construct very-tall buildings and promote commercial development in specific zones, which would depend on effective and appropriate technology and resources.

The urbanisation processes are largely driven by market forces and government policies that lead to simultaneous processes of change in livelihoods, land-use, health and natural resources management including water, soil and forests and often reactive changes in local governance. Government development policies and budget allocations, which often favour urban residents over rural areas, tend to pull people into the urban areas. In the cities, public investment, which often misses the urban poor, with expenditures biased towards the higher income classes and poverty among vulnerable groups such as new migrants force them into slums and squatter settlements.

The market forces result in a series of changes in employment, urban agriculture and semi-urban production systems. Ready market leads to an increase in production of horticultural crops and perishables such as vegetables given the high demand and proximity to urban consumers. These factors, therefore, act as a great centripetal force in favour of urbanisation.

Current Urban Challenges– An Overview

Urbanisation leads to the outward expansion of cities and results in changes in land-use whereby urban residents buy up prime agricultural land for residential or commercial purposes. The conversion of farm lands and watersheds for residential purposes have negative consequences on food security, water supply as well as the health of the people, both in the cities and in the semi-urban areas.

Cities impact on health in many ways. In the areas of the environment and health, problems of emission reduction, supply of clean drinking water, sewage and rubbish disposal, food security and poverty reduction are the most important. Vulnerability of the urban population to natural disasters and diseases, especially HIV/AIDS and atmospheric pollution has also been recognized. Although, data about pollution levels are fragmentary, the air and water quality in many cities threatens the health of millions of city residents (UNEP, 1994). Although, a significant positive impact of urbanisation is promotion of urban agriculture and the cultivation of staple crops, vegetables, poultry and dairying, which are demanded by urban consumers, cultivation of vegetables through sewage irrigation and the use of chemical pesticides affect the health of consumers who are not notified of the circumstances of cultivation of these products.

A large proportion of the urban population is also affected by poor sanitation that threatens their health. River pollution is particularly found to be worse where rivers pass through cities and the most widespread is contamination from human excreta, sewage and oxygen loss. It is estimated that about 400 million people or about one-third of the population in the developing countries do not have safe drinking water (World Bank, 1990). In many areas of Imphal East and Imphal West, there is limited access to clean drinking water. The quality of several watercourses is poor, with pollutant levels higher than the WHO's standards. Pesticide contamination from urban agriculture, residues from sawmills and manufacturing industries, wastewater from urban drains and municipal dumping of waste especially human excreta pollute drinking water sources that affect the health of the urban and semi-urban populations. In the long-term, treatment of sewage would be required for safer vegetable production and to reduce water pollution.

Urban populations are also vulnerable to diseases such as malaria or those associated with air pollution. Other malfunctions that are associated with industrial and traffic injuries and psychological disorders, especially in low-income urban and semi-urban areas are also disturbing. The unhealthy environment and overcrowded housing in the urban expose the urban poor to high rates of infectious diseases such as pneumonia, tuberculosis and diarrhoea. Although it is clear that cities in the developing countries act as nodes through which development occur, it is important to note that rapid urbanisation poses particular risks that affect sustainable livelihoods of millions of people. The wide range of effects includes degradation of the environment (soil erosion, deforestation), destruction of watersheds and wetlands, traffic congestion, contamination/pollution of water, and environmental risks associated with low-income housing areas.

Effects of Population Growth and Urbanisation in Manipur

More than 5,75,968 people of Manipur live and work in towns, and the rate of urban population growth throughout most of the State is 25 per cent. Urbanisation, no doubt, has significantly improved the economic prospects and quality of life for a large and increasing proportion of the people of the State. However, as this section shows, it has also caused many problems.

- **Population Growth and Migration:** The effect that high population growth rates have had in slowing economic growth appears to be not well understood by policymakers. Key drivers of these trends include push factors, such as declining agricultural commodity prices and livelihood opportunities and insufficient rural land to confer social standing, as well as pull factors, such as the prospect of cash employment, perhaps with the government, the availability of public services in town and the intrinsic excitement of urban areas.
- **Urban Unemployment:** The State of Manipur has, so far, not enjoyed the dramatic private sectorled economic growth of other States of India. Indeed, economic indicators in many countries have been stagnant—or even negative—for some time.
- **Social Problems:** Urbanisation is also closely associated with social problems like, unemployment, slum development, poverty and environmental degradation. HIV/AIDS were quite prevalent in the urban areas of Manipur.
- **Health Problems:** The most prominent public health problems in the State, especially among the poor, remain those of (largely preventable) infectious diseases, in particular respiratory diseases related to overcrowding, and gastro-enteric diseases related to water pollution, poor sanitation,

and inappropriate health and hygiene practices. Poor-quality water and sanitation services continue to contribute to public health risks. High fever and infant diarrhoea are common throughout the region.

- **Lack of Adequate Education:** Notwithstanding a policy orientation in many States of India that seeks to minimize or even reverse urbanisation, education systems in the State are orienting young people to non-manual employment in towns, primarily in government. It has not equipped them with the technical and business skills demanded by the emerging urban population.
- **Pressure on Infrastructure and Services:** While economic prospects in urban areas remain significantly better than in rural areas, the productivity of the urban economy and quality of life in Manipur is deteriorating, particularly for the poor, because of deficiencies in essential infrastructure, utility services, and housing. In many towns these services were established before independence, and increasing demand from the rapidly growing urban population has overwhelmed them.
- **Water Supply:** The problems associated with delivering satisfactory water supply in urban areas are very poor. They reflect inappropriate policies, undue government interference, and the lack of appropriate incentives for consumers to reduce demand to sustainable levels, all of which undermine the ability to operate and maintain water supply systems properly.
- **Solid Waste Management:** The institutional framework for disposal of solid and industrial (including hazardous) waste in Manipur towns is outdated and ineffective. In most of the towns, solid waste collection is a local government responsibility and accounts for a significant proportion of local budgets. Without exception, these systems lack sufficient financial resources and technical expertise. As a result, even where the percentage of waste collected is high, the maintenance of equipment and facilities is poor and long-term sustainability is in doubt. Affordability and willingness to pay are questionable, particularly for low-income households, and cost recovery is low. Even more serious problems relate to disposal.

Imphal Municipal garbage dumps are usually located in Lamphelpat area. They are inadequately sealed from flooding, creating a potent environmental hazard throughout their lifespan. Most agencies responsible for waste disposal are unclear as to the nature and quantities of the waste they handle. The municipality has also introduced innovative programmes to involve unemployed youth in city cleanliness programmes. However, the municipality lacks the financial and technical resources to maintain and replace collection vehicles, equipment, and facilities, much less promote waste minimization and recycling or formulate and enforce policies relating to disposal of toxic wastes. Poor drainage disrupts economic activity; compounds the problems of poor housing, particularly for low income families; and increases health risks throughout the State.

- **Power Supply and Telecommunications:** The growing interest in commercializing utility operations in the State is perhaps most evident in the power and telecommunications sectors. In those sectors, reliable and competitively priced power and telecommunications services are essential to attracting the investment necessary for growth in urban economic activity and employment, particularly in new higher value-added activities in information industries. But, load shedding of power supply in the State is very common.

Conclusion

In Manipur, urban areas are expanding rapidly. For the majority of urban dwellers, especially the poor, finding potable water supply, affordable shelter, accessible and secure urban land for agriculture to ensure food security, securing gainful employment and improvement in health facilities would continue to remain a priority. Since restrictive urban growth policies, especially population distribution designed to reduce the rate of rural-urban migration appear to have had limited success in many developing countries, policies must be directed at transforming the rural economy in order to slow the rate of urban sprawl. Comprehensive land-use planning and revision of planning standards and administrative procedures would, go a long way to reduce many of the problems that face urban population in the State.

On the other hand, after examining the growth of population in Manipur, two main responsible factors emerge. One is the high birth rate and declining death rate and the other is migration. The factor of high birth rate and low death rates may be considered as a natural demographic transition. But, the second factor of migration alarms the general public at the highest level. Migrants constituted about 6.00 per cent of the total increase in population during 1961-71 in which 17,667 migrants were added to the population of the State. It was 20,399 during 1971-1981 (Economic Review of Manipur, 1994-1995). Therefore, it can be concluded that migration from other States of the country and immigration from other neighbouring countries can significantly contributed to the growth of population of this State. On the other hand, the uneven growth of population among the different groups of ethnic religious communities needs the urgent attention of all. This may be due to the feeling that increase in population in their own groups might be assets to their own groups or communities.

Whatever may be the causes of growth of population, the growth rate itself has been very rapid since1951 after it has been merged with the Indian Union in 1949. The more alarming point is that the trend of population growth was significantly much above the national averages during the last 6 (six) decades. Further, the distribution and density of population between the hill and the valley had shown a great difference. Though density of population in the State is not so high in comparison with the national average, the heavy concentration of the population in the small valley of the State is a matter of great concern. For example, as per the 2011 Census, the density of population in the State was 122 persons per sq. km. It was 733 in the valley and 57 in the hill regions. The national average was 368 at the same period. The relationship between population growth and economic development of a State has been the focal attention of the whole world. It may be desirable for a particular State or country to obtain growth of population upto a certain level because human beings are both means and ends of all economic activities. However, rapid growth of population will become a liability and extra burden to the State or country. And return, it affects economic development of a country. Problems of unemployment, poverty, malnourishment, etc., are all the example of evils of population explosion. It is worthwhile to mention here that unemployment levels reached 6,40,010 persons in Manipur as on 31st July, 2008 (Employment Exchange, Imphal).

It was 5,83,653 persons as on 31st March, 2007. The rate of increase per month was 1,177 persons. Out of the total, 4,21,623 were male and 1,63,207 were female candidates searching for jobs. In this regard, a proper policy of population needs to be formulated in order to meet the challenging issues of economic development of the State.

Last but not the least, the pressure of population in the valley areas is very high, particularly in the capital town of Imphal. Imphal West District continues to be the most populous district of Manipur. There are 514,683 persons or 18.91 per cent of the total population of the State in Imphal West district alone (2011 Census). Density-wise also, Imphal West District has the highest density (992 persons per sq. km) followed by Thoubal District (818 persons per sq. km.) and Imphal East District (638 persons per sq. km.). On the other hand, the most populated hill district i.e. Senapati has a population of 354,972 persons or 13.04 per cent of the total population of the State (2011 Census). The average density of the total hill region was 57 persons per sq. km against 733 persons per sq. km of the total valley region (2011 Census). Therefore, the extension of the Manipur Land Revenue and Land Reforms Act 1960 with necessary modifications in the hill areas of the State is the need of the hour for proper utilization of land areas and equitable distribution of economic and sustainable development of the State. It is pertinent to mention here that the absence of the said Act of 1960 restricts the valley dwellers to move into the hills and settle there (which can otherwise be utilised for economic activity), whereas anybody from any parts of the country (whether from hills or plains) can come and settle in the valley of the State and thereby increasing the pace of urbanization in the valley without any corresponding increase in industrialisation in the State.

NOTES

1. Jansankhya Sthirata Kosh (National Population Stabilisation Fund), Poverty and Population, http://www.jsk.gov.in/poverty_and_population.asp accessed on 17th March, 2011.

REFERENCES

Angotti, T. (1993). *Metropolis 2000: Planning, Poverty and Politics*. London: Routledge.

Canales, A. (1999). Industrialization, Urbanisation, and Population Growth on the Border. *Borderlines*, Vol. 7 (7).

Directorate of Economics and Statistics (1992). *Economic Review of Manipur: 1990-1991.* Imphal: Govt. of Manipur.

Directorate of Economics and Statistics (2001). *Economic Review of Manipur: 1999-2000.* Imphal: Govt. of Manipur.

Dun, E.W. (1975). *Gazetteer of Manipur.* Imphal: Govt. of Manipur.

Government of Manipur (2009). *Statistical Abstract of Manipur.* Imphal: Directorate of Eco. & Statistics.

Henderson, J.V. (2004), Urbanisation and Growth, in Brown P. Aghion and S. Durlauf (eds.), *Handbook of Economic Growth*, Volume 1, University Press: North Holland.

Kundu, A. (2006). Trends and Patterns of Urbanisation and their Economic Implications in Govt. of India (Eds.) *India Infrastructure Report.* New Delhi: Govt. of India.

Kwasi Nsiah-Gyabaah (2002). Urbanisation Processes—Environmental and Health Effects in Africa, Panel Contribution to the PERN Cyber Seminar on *Urban Spatial Expansion PERN*; Steering Committee member and Principal, Sunyani Polytechnic.

Mahendra K. Premi (1991), *Indian Population—Heading towards a Billion.* New Delhi: B.R. Publishing Corporation.

Raskin, P., G. Gallopý'n, P. Gutman, A. Hammond, and R. Swart, (1998). *Bending the Curve: Toward Global Sustainability.* Stockholm: Stockholm Environment Institute.

Singh, Nixon and Prasain, G.P. (2010). Population Growth in Manipur: A Demographer's View, in Bimal J. Deb (Eds) *Population and Development in North East India.* New Delhi: Concept Publishing Company Pvt. Ltd.

UNCED (1992). *Agenda 21 of the Earth Summit, the Rio Declaration on Environment and Development.* Rio de Janeiro: UNCED.

UNDP (1996). *A Survey of All Urban Areas of Fiji.* Fiji: Government of Fiji/ UNDP. Retrieved on August 5, 2009 from www.siteresources.worldbank.org

UNEP (1999). *Global Environment Outlook.* London: Earth Scan Publications Ltd.

UNFPA (1993). *The State of the World.* New Yok: UNFPA

United Nations (1991). *World Urbanisation Prospects*, New York: UN.

United Nations (UN) (2002), Press Release, POP/815, 21 March, p 1.

United Nations Development Program (1990). *Human Development Report*. New York: Oxford University Press.

United Nations Environment Programme (UNEP) (1994). Air Pollution in the World's Mega-cities. *Environment*, Vol. 36 (2), pp. 4.

UNPD (2001). *World Urbanization Prospects: The 1999 Revision. Key Findings*. New York: United Nations Population Division. Retrievedon September 12, 2011 from www.un.org/esa/population/pubsarchive/urbanization/urbanization.pdf [Geo-2-203].

World Bank (1990). *Annual Review of Urban Development in the Third World.* Washington D.C.: World Bank.

World Bank (1999). *Entering 21st Century, World Development Report 1999/2000*. New York: Oxford University Press.

13

Unstinted Jaintia Hills of Meghalaya: An Unnoticed Reality from Development-Lens

Amitabha Ray

Meghalaya attracts nature lovers and globe trotters because of its picturesque natural beauty. Tourists from different parts of India and other countries regularly visit Meghalaya to enjoy the serenity of natural beauty of Shillong, Sohra (Cherrapunjee), Elephant Falls, Nohkalikai Falls and a couple of other falls. The hot springs at Jakrem near Mawsynram are believed to have curative and medicinal properties. The subtropical forests of Meghalaya support a vast variety of flora and fauna. Two National Parks and three Wildlife Sanctuaries are added attractions. Meghalaya also has many adventurous tourism opportunities in the form of mountaineering, rock climbing, trekking and hiking, water sports, etc. The State offers several trekking routes which unfolds opportunities to have a glimpse of some rare animals such as the slow loris, assorted deer and bear. The Umiam Lake has a water sports complex with facilities such as rowboats, paddleboats, sailing boats, cruise-boats, water-scooters and speedboats. Even one may prefer not to miss the recently modernised age old Mawsmai Cave near Sohra.

Incidentally, most of the tourists do not show any interest in making a trip to Jaintia Hills. Only the commuters to Badarpur and Silchar take the road bound for Jowai, which is the administrative headquarters of Jaintia Hills District as well as the commercial centre.

In fact, Jowai is only 64 km from Shillong and is situated alongside the Myntdu river. You will definitely enjoy Nartiang. The village of Nartiang is equally a point of attraction because of the Summer Palace of the Jaintia King whose Capital was Jaintiapur. Nartiang Monoliths is the biggest collection of monoliths or megalithic stones in one single area is to be found north of the Nartiang market. Thadlaskein Lake, Umhang Lake, Kiang Nongbah Monument, Stone Bridge at Thlumuwi, Syntu Ksiar (Flower of Gold), Umlawan Cave, Ialong Park, etc., are also worth visiting But your visit to Jaintia hills is incomplete till you have an experience of rat-hole mining.

Why Jaintia

Jaintia Hills District is one of seven districts in Meghalaya situated in the east of the region, with Bangladesh to the south and the State of Assam to the north and the west. It covers an area of 3,819 sq. km. which is 17.03 per cent of the total geographical area of the State. According to the Census of India (2011), the district, with a population of 392,851, is similar in its societal structure to Meghalaya as a whole. It is observed that there was an increase of 31.34 per cent compared to 2001 Census (299,108). As a result, Jaintia Hills district contributes 13.25 per cent of the entire population of the State. Around 93 per cent people live in rural areas where literacy rate is below 61 per cent. On the other hand 7 per cent people live in urban regions where literacy rate is about 92 per cent. It shows that almost 96 per cent of the population belong to the Scheduled Tribes. The district has retained an indigenous administrative system which still works as an authoritative governing structure. Approximately 49 per cent of the district population live under the poverty line. Majority of the working population (75.7%) is involved in agriculture.

The Meghalaya region is rich in natural and mineral resources such as coal, limestone, fireclay and phosphorite. According to the Directorate of Mineral Resources Meghalaya the State is rich in tertiary coal deposits with an estimated 640 million tonnes in reserves. About 40 million tonnes of which is in the Jaintia Hills District. Between 1997 and 2003, coal production in Meghalaya increased by 36 per cent, and in 2002-03 coal production in Meghalaya accounted for approximately 0.7 per cent of that of the whole country. The mining industry contributes 8-10 per cent of GDP in the State of Meghalaya.

Jaintia Hills District of Meghalaya is a major coal producing area with an estimated coal reserve of about 40 million tonnes. There are 9 important coal deposits, out of which Bapung and Lakadong are the most important, the others being Lumshnong, Malwar Musiang Lamare, Mutang, Sutnga, Jarain Tkentalang, Ioksi and Khliehriat. Out of the coal bearing areas are accessible by roads from Jowai. The NH-44 connecting Shillong, Jowai, Badarpur and Silchar leads access to Bapung, Malwar, Mutang, Lumshnong and Lakadong coalfields. The coal bearing areas of the district present a panorama of flat topped low hills, devoid of vegetation and plateau of rolling grasslands intersected by river valleys.

The coal, in the area is found imbedded in sedimentary rocks, sandstones and shale of the Eocene age. Three coal seams vary from 30 to 212 cm in thickness. The main characteristics of the coal found in Jaintia Hills are its low ash content, high volatile matter, high calorific value and comparatively high sulphur content. The coal is mostly sub-bituminous in character. The physical properties typify the coal of Jaintia Hills District as hard, lumpy bright and jointed except for the coal in Jarain which is both soft and hard in nature. Composition of the coal revealed by chemical analysis indicates moisture content between 0.4 per cent to 9.2 per cent, ash content between 1.3 per cent to 24.7 per cent, and sulphur content between 2.7 per cent to 5.0 per cent. The calorific value ranges from 5,694 to 8,230 Kilo-Calories/Kilogram.

The existence of rich coal and limestone deposits in Jaintia Hills is a great boon to those who wish to start any industrial venture based on these two minerals such as the manufacture of Portland cement. Besides coal and limestone, availability of other minerals like clay, shale, phosphate have also been confirmed in the Jaintia Hills.

Ownership of the Mines

Coal mining in particular has been a growing industry since the Supreme Court's ban on the sale of timber for commercial purposes in 1981, which had been an important source of income for many farmers. Mining began in the State of Meghalaya over 30 years ago and has since turned into one of its most profitable markets, showing no signs of slowing down. Under the sixth Schedule of the Constitution of India, the tribal and native communities residing in Meghalaya have exclusive rights over their land which include for mining operations. More than 70 per cent of its land is used in mining activities and large scale extraction is carried out which *prima facie* in the absence of government regulation. The number of mines in the district is uncertain due to unenforced government registration and a lack of interest in conducting this kind of count. Local people, however, estimate that the number of mines is into the thousands.

Law of the land does not prohibit unscientific mining. The owners of the property consider that they have right to use the land for whatever purpose. As an area that is dominated by tribal groups, ownership of the land is handed down by elders to individuals, and occasionally to communities or villages. This system has encouraged mushrooming of privately controlled small-scale ventures to exploit the s surface level coal in the region.

As already indicated, as per provision of the Sixth Schedule of the Constitution of India, the land is solely owned by the people and the State and Centre have little or no control whatsoever. Consequently, these mines are named as local cottage mines or small-scale coal mines, operate outside the purview of Coal Mines (Nationalisation) Act. It is learnt that Government has framed a draft mining policy to restrict the unscientific coal mining practices. The proposed policy prescribes series of measures to promote scientific mining and underscores an urgent need to see that the environment is protected while initiating various mining activities. Since, the proposed policy is yet to be enacted, the unscientific mining practices by individuals/community/village as owner of land remain unabated. This is largely known as rat-hole mining.

Rat-hole Mining

It is a primitive and traditional mining process, adopted locally for extraction of coal. The hole in mine area has a one square metre opening along which the miners crawl and excavate coal. The miners go as deep as 50-100 metres in length from the opening. In this method the land is, first cleared by cutting and removing the ground vegetation and then pits ranging from 5 to 100 square metres are dug vertically into the ground to reach the coal seam. Thereafter, horizontal tunnels are made into the seam for extraction of coal. Extracted coal slabs/ pieces are brought to the bottom of the pit by using a conical basket or a wheel barrow.

The horizontal rat-holes are approximately a kilometre long and only has height enough for children to crouch in whilst working. The holes wind around like a network of mole burrows, and at various points they connect with other rat-holes from other vertical access shafts. There are no measures in place to protect against cave-ins, and no lighting is provided.

A minimum of 25 metres is dug to extract coal. The labourers, including children, use wooden steps inside the mine and remove coal. The coal is taken out of the pit and dumped on nearby un-mined areas, from where it is carried to the larger dumping yards near highways for onward transportation. The present coal exploitation, reportedly, is about 2 million tonnes per annum in this district. Most of the mining activities are small scale ventures controlled by individuals who own the land.

The entire process of mining is done manually employing small implements. Even, excavated coal is manually carried to the roadside for loading onto trucks. Without access to expensive or sophisticated machinery, landowners have adopted a crude form of excavation that employs rat-hole miners. The miners carry into the mine a small pickaxe and a cart to load the coal into, and they work at the coal face. When they have mined enough coal to fill the cart, they drag it back to the base of the access shaft. Without the use of appropriate facilities, a make-shift arrangement made of bamboo and wooden pulley collects coal into metal boxes and a crane takes them to the surface. If there is no crane, the coal is put into a bamboo-weaved basket and carried up the stairs to the surface on a worker's back.

The mines are operated with primitive practice without bothering for compliance to the minimal safety standards. Safety training or equipment for workers are distant dreams. As the Meghalaya region has high level of rainfall, the wooden steps used for reaching bottom of the shaft are often slippery and prone to rotting. Slips, falls, fatal injuries are not uncommon.

Child Labours

As in the rat-hole mining one has to enter mine through only 1 square metre opening and there is no other option left other than crawling. So, children are considered as the best miners here. It is told that under life-threatening conditions, an estimated 70,000 children work in the coal mines in the Jaintia Hills, according to a children's rights organization working to end the practice. The youngest of the miners are just 7 years old.

You may learn that for the equivalent of a few dollars a day ($ 5) per cartload of coal they work in narrow, unreinforced seams in small mines. Veterans of the locality may inform you without disclosing their identity that most of the children are illegally brought here from the neighbouring countries mainly from Bangladesh and Nepal. The Jaintia Hills district administration conducted a sample survey in 20 villages in July 2009 and found that there were 222 children working in the coal mines. While 153 were local residents, the remaining were from Assam, Bihar, Nepal and Bangladesh. Some have been sold by their families as indentured labourers.

It is told that on an average, a labourer earns about Rs. 400 ($ 8) a day. Those who extract coal from the 'rat-hole' get about Rs. 1500 ($ 30) a day but they risk life. Inside the mine, life is very tough. It is humid and very hot. One cannot work for more than four hours at a stretch inside the mine and it causes suffocation. Children can work 2 and a half hours. After that they have to come out for fresh oxygen. Children are usually trained informally by seniors or relatives for 10-15 days before they start extracting coal by themselves.

Economics of Rat-hole Mining

Most of the output of these mines, reportedly, is dispatched sometimes through illegal route, to markets in Bangladesh, without any local value-addition or use in industrial activities. There is very miniscule growth of industries that can use coal, neither there is presence of any coal-based industry in the region. Still, the coal mining industry boosts the local economy, increases employment, and enhances the incomes of the local population.

Mines with larger size normally employ 200 labourers, while small mines have not more than 35 to 40 labourers. It is an open secret that some managers hire villagers, paying them a small commission to deploy labourers to work in their mines for six-month terms. It is also known that there are brokers who recruit labourers, including child labourers.

The labourers for extracting, breaking, sorting, loading and unloading coal is organized into groups with a leader called a Sardar. This person is usually of non-tribal origin, often Nepalese, and sometimes a woman. (S)he is paid a percentage for every truck loaded, taking Rs. 200 ($ 4) out of the approximate Rs. 1,200 ($ 24) per truck. When the mine is small, the Sardar himself/herself also joins the workforce; otherwise (s)he supervises the other workers. At times, Sardars also provide mine owners some of the tools needed to extract coal against charges. The individual miners are reported to earn Rs. 800 ($ 16) for each 5 square metres coal tray filled.

By knocking the doors of different stakeholders if you are able to generate their confidence, you will learn that majority of the output is sent to the port of Chittagong, the largest seaport in Bangladesh, and shipped from there to foreign countries. Yet, there is no reliable data or evidence to estimate the quantum of drainage of this scarce resources of high quality.

Jaintia society follows the matriarchal tradition where the youngest daughter gets ownership of the family home, land and other property. So, many of the mine owners are women. Described as 'highway queens', these young women can be seen keeping a watch on their stockpile of coal along the National Highway and overseeing its loading and sale. Another group is seen sitting with large, wooden measuring boxes, buying coal heaped in overloaded trucks at the local retail price and then selling it back to the truck driver beyond the weighbridge, of course, at a profit.

Safety and Living Condition

The working conditions at the mines are extremely hazardous. The ladders that labourers use in order to go down to the mines are incredibly dangerous, especially in rainy season. The children are sent deep into the rat-holes to extract coal, where they work for many hours at a time. At times there is little oxygen as the holes are extremely deep and there is insufficient ventilation. The general working environment is extremely dangerous because of primitive technology. There is a total lack of safety measures to prevent accidents, which occur frequently and are often fatal. No equipment, apart from headlamps, is provided to protect labourers from accidents. Frequent flooding due to unplanned digging by labourers for constructing a new hole sometimes cause the fatal accidents. It is learnt that in 2009, unscientific digging by labourers pierced a river-bed and the consequent flooding killed more than 80 people, including children.

If you are able to develop rapport with any mine manager, you will come to know that there is no money kept for safety measures. Cave-ins are always a threat; wooden ladders leading down to quarries are slippery with moss; there is little or no access to medical care, sanitation, safe drinking water, or even adequate ventilation. On an average, there are eight accidents a month in the mines, at least two of which are fatal. If any mine owner is asked about the dismal working conditions and child labour, they claim as media rumours.

One can observe that labourers of the mines, including children, are living in inhuman conditions. There is no access to safe drinking water. Sewage system does not exist and as a result residents suffer from various diseases. In spite of these, mine owners provide no medical facilities for labourers.

When the entire system is operated without any articulated and structured institutional mechanism one should not search for any occupational safety regulations and labour organisation at the mines. Naturally, the mine owners do not abide by any form of safety procedures or labour standards. In addition, there are no workers' unions or any other form of organization among the labourers that could enable them to raise grievances with management. In this unregulated and unsupportive work environment, accidents are frequent and mine owners continually fail to improve safety and respond adequately after accidents happen. Since the mines in Meghalaya are built by manual labourers who usually have no professional engineering expertise, they are very fragile and liable to collapse. Ceilings of the tunnels often cave in, and steps and bridges inside the mines break. Water build-up in mine cavities is a constant and deadly threat to labourers. The build-up creates pressure on mine walls, causing stones to fall on and crush workers. In addition, the pressure can break the walls, flooding mine cavities and drowning labourers.

To illustrate the serious dangers that miners face, the local people, without disclosing their names described an accident at a mine named *Setu* which had occurred in 2009. In that accident, 81 labourers were killed when subterranean water from a nearby river flooded a mine cavity. Since no warning system existed at the mine, none of the victims had time to escape. Following the accident, the mine owner failed to drain the accident site and remove the victims' bodies. No monetary compensation was provided to the victims' families.

Mine managers and owners commonly keep a register of their labourers' identities and wages. Since these registers do not typically include home addresses or family information, however, contacting family members in the event of the injury or death of a labourer is difficult.

Labourers' safety and health concerns extend far beyond mine accidents. It is learnt that physical abuse by mine owners and managers against labourers. Some owners and managers corporeally punish labourers, most frequently for committing theft, for fighting, and in retaliation for unpaid debts. In addition, it is reported that trafficking for prostitution of children in the areas surrounding the mines is a regular feature. Local authorities often fail to investigate and prosecute these serious rights abuses.

Is it possible to expect adequate medical care and nutrition for labourers in this area? The medicines labourers use to treat workplace injuries or illnesses frequently have expired validity dates. Furthermore, labourers often take the wrong medicine after seeking treatment advice from unlicensed pharmacists. As a result, many labourers who seek treatment become more ill or even die. In addition, some labourers fall ill or die after eating spoiled or improperly cooked food.

Although accidents often cause deaths and injuries, incidents are not reported to the police and mine owners provide no compensation or medical expenses for victims or their families. Mine managers do not have record labourers' home address or family details, so when a labourer dies families are often not informed of their loss. Unclaimed bodies are cremated and buried around the mines. The deaths of missing children go unreported.

Effect on Environment

Ecosystem disturbance is an event or series of events that alters the relationship of organisms and their habitat in time and space. Ecosystem disturbance caused due to mining is an evitable fallout of industrialization and modern civilization. Forest ecosystems have important functions from an ecological perspective and provide services that are essential to maintain the life-support system on a local and global scale. The rate and the intensity of land-use and land cover change are very high in developing countries because of various human activities. Mining of coal both surface and subsurface causes enormous damage to the flora, fauna, hydrological relations and soil biological properties of the systems. Destruction of forests during mining operation is invariably accompanied by an extensive damage and loss to the system. The overburden of coal mines when dumped in unmined areas creates mine spoils which ultimately affects the surrounding vegetation.

The extraction of coal creates a variety of impacts on the environment before, during and after the mining operations. The extent and nature of impact can range from minimal to significant depending on a range of factors associated with ongoing mining processes as well as post-mining management of the affected landscapes. The sensitivity of the local environment also determines the magnitude of the problem. Usually, an ecologically fragile environment has been found highly vulnerable, attracting long term ecological impacts.

Mining operation in Jaintia Hills undoubtedly has brought wealth and employment opportunities in the area, but simultaneously has led to extensive environmental degradation. Large scale denudation of forest cover, scarcity of water, pollution of air, water and soil and degradation of agricultural lands, deforestation, water pollution, land subsidence, dust pollution and increase in wasteland, acid mine discharge, changes in the land-use patterns of the area are some of the conspicuous environmental implications of coal mining. Besides, caving in of the ground and subsidence of earth and haphazard dumping of coal an overburden have deteriorated the aesthetic beauty of the landscape.

The waterbodies of the area are the greatest victims of the coal mining. The problems of water quality degradation and its adverse impacts on availability of potable and irrigation water, soil quality and agricultural productivity, and biodiversity in the area have been attracting increasing attention of people.

According to reports, coal mining in Meghalaya popularly known as "rat-hole mining" has changed the ecology and landscape of the hill State, besides affecting the health of the migrant children and their parents working in the mines.

An Open Secret

You will be astonished to notice that starting from stakeholders of the rat-hole mining to the administration, everybody is well aware of this inhuman state of affairs. Even they are ready to informally speak out in this regard. But, most of the people are afraid to express anything on record. Elite international media channels have several times telecasted this inhuman practice; all national and internationally famous newspapers have frequently brought out stories on rat-hole mining of Jaintia Hills. You can recall that in most the cases either the names are changed or the faces/figures are blurred.

Only exception is Ms. Hasina Kharbhih. This young Khasi lady, a pass-out of IIM-Ahmedabad and head of a non-governmental organization, is the one of the leading activists engaged in wiping out such an inhuman business. She has boldly brought out the issue in different platforms whether it is a national one or an international forum. It is said that due to continuous pressure from the administration and some non-governmental organizations some of the mine owners have stopped hiring children.

Child Rights and Trafficking

Based, *inter alia*, on complaints received about large scale child labour in the coal mines area of Lad Rymbai and the alleged transfer of children from Jaintia Hills, Meghalaya the National Commission for Protection of Child Rights ('NCPCR or Commission') team visited Meghalaya from August 8th to 11th, 2009. The NCPCR observed that due to growth of industry and township around it, there has been a phenomenal demand for cheap labour in Jaintia Hills. As it is well established, children are preferred because they are the cheapest form of labour and can be forced to work in inhuman conditions. Finally the NCPCR suggested that there has to be a sustained and massive campaign against all forms of child labour and a generation of public support for implementation of all the relevant laws.

The NCPCR also proposed that the labour department must enforce the Child Labour (Prohibition and Regulation) Act 1986, Juvenile Justice Act as well as the Bonded Labour System Abolition Act (1976). strictly and register cases against employers of children in domestic work, hotels, eateries and other establishments, mining and other sectors where child labour is prohibited and coordinate with the police and revenue departments for enforcement of the laws.

The strategic location of Jaintia Hills on the largely porous border of Bangladesh facilitates the easy export of coal to the international market and the infiltration of cheap labour into the coal industry. The absence of State regulation of the coal industry and the lack of strict enforcement of labour laws has encouraged local entrepreneurs to exploit the cheapest sources of labour that would ensure maximum profit—poor migrant children from Nepal, Bangladesh and from other

States of India have, therefore, become a natural target for the coal entrepreneurs in the Jaintia Hills. The unfortunate prevalence of child labour in the context of human trafficking that constitutes severe violation of international conventions on human rights and children's rights has drawn the attention of the international community to the plight of children illegally employed in the coal mines in Meghalaya.

Sporadic newspaper reports estimate that there are 10,000 people, mostly children trafficked every year into North-East India from Nepal, Bangladesh and South-East Asian countries, there is an absence of systematic research to estimate the specific number of children working in the coal industry in the Jaintia Hills. Preliminary investigations by Impulse NGO Network since March 1997 has confirmed the presence of a large number of children illegally employed in the coal industry in Jaintia Hills. A Baseline Field Study on the Survey in Jaintia Hills District, Meghalaya, was conducted by interns from Tata Institute of Social Sciences, Mumbai from March 28, 2007 to April 3, 2007 under the supervision of Impulse NGO Network.

Need of the Hour

Thus, the situation in Jaintia Hills of using child labour under 14 years old constitutes a violation of Article 24 (Prohibition of Child Labour) of the Constitution and Child Labour (Prohibition and Prevention) Act of 1986. Other constitutional violations include (i) Article 38 (State to promote welfare of the people), (ii) Article 39 (Health of Children not to be Abused; Children not to be forced into Unsuitable Work by Economic Necessity), (iii) Article 41 (Right to Education), (iv) Article 45 (Free Education), and (v) Article 47 (Duty to Improve Health).

Moreover, the situation in Jaintia Hills using children under 18 years old constitutes a violation of (i) Mines Act of 1952 (Updated to 1983) as well as (ii) Mines Rules of 1955. As discussed above, the mines also rampantly violate safety and other rules for the protection of workers under (i) Mine Rules of 1955, and (ii) Coal Mine Regulations of 1957, which violations especially negatively impact child labourers.

Ray of Hope

It is learnt that Meghalaya Government is working on a rehabilitation package to help the children working under hazardous situations in the coal mines of Jaintia Hills, following a direction from the National Human Rights Commission. Recently Government of Meghalaya declared that coal mining activities would fall under the National Coal Mining (Conservation and Development) Act 1974. Till then, Meghalaya was the only State in the country where miners were not required to take a mining lease from the government. Now, the State government has engaged the Coal India Ltd (CIL) to prepare a plan for institutional mining in the region and further decisions on the matter would be taken accordingly.

Conclusion

The land ownership system in the State needs to be reviewed and implementation of mine regulations is also to be followed strictly to reclaim the mined lands to an ecologically improved and more socially approved conditions. In abandoned mine areas afforestation programmes the need to be taken up by mine owners. The rat-hole mining should be discouraged and alternative scientific methods should be promoted to extract coal. The reclamation of mining areas should be done. The 'acid mine discharge' is to be diverted away from potable water sources. Proper health care facilities need to be provided to labourers working in the mines.

Refilling of the mined areas with mine spoils/top soils followed by a re-vegetating scheme with acid-tolerant species including leguminous plants can be carried out. The Meghalaya State Pollution Control Board should also be given the tasks of periodic analysis of water, soil, and air from the mining areas. Local leaders too should be educated on the task of restoration of the mined areas.

To summarise a holistic approach by all stakeholders, i.e., NGO's, ecologists and government agencies is needed for the eco-restoration of mined areas, introduction of modern mining practices and finally cessation of child labour. It would be prudent to utilise a part of the revenue earned from coal mining for such development programmes.

REFERENCES

Government of India (2011). *Census of India* 2011. New Delhi: Office of the Registrar General & Census Commissioner.

Higashi-Ueno, Taitou-ku (2011). *Report on Child Labour in the Mines of Meghalaya.* Tokyo, Japan: Human Rights Now. [Retrieved on August 2, 2011 from http://www.hrn.or.jp/eng]

Impulse NGO Network (2010). *An Exploratory Study of Children engaged in Rat-Hole Mining in the Coal Mines of Jaintia Hills District, Meghalaya.* Shillong, Meghalaya: Impulse NGO Network

Impulse NGO Network (2010). *Investigation of Child Trafficking in the Coal Mines of Jaintia Hills District, Meghalaya.* Shillong, Meghalaya: Impulse NGO Network

14

Mass Media Intervention and Health Sector: Tales of Maternal and Child Health from Assam

Silajit Guha and *Biraj Kanti Shome*

Introduction

The State of Assam is the most populous, important and the biggest State in the North-Eastern region of India with a population of 31.16 million *(Census 2011)*. The State is almost isolated from the rest of the country except from a narrow corridor known as chicken's neck running through the foothills of Himalayas. It is bordered in the North and East by the Kingdom of Bhutan and Arunachal Pradesh; along the south, it has Nagaland, Manipur and Mizoram. Meghalaya, West Bengal and Bangladesh are her Western neighbours. Its geographical area is 78,438 sq km with a population density of 397 per sq km (as per *Census 2011*). 87 per cent of the State's population resides in the rural areas and remaining 13 per cent in urban areas. The State exhibits diverse geographical contours and largely dependent on an agrarian economy (Table 14.1).

It is also the State which has been identified by the Government of India as one of the 18 High Focus States (HFS). This categorization is based on the weak public health indicators and infrastructure. For example, maternal mortality ratio is 381 (Annual Health Survey 2010-11) per 100,000 live births, which is the highest in the country.

Table 14.1: Demographic Trends of Assam as compared to India

	Assam	*India*
Total Population (in millions)[a]	31.16	1210.19
Sex ratio (females/1000 males)[a]	954.00	940.00
Crude Birth Rate (per 1000 population)[b]	23.20	22.10
Crude Death Rate (per 1000 population)[b]	8.20	7.20

Source: [a]Census 2011; [b]SRS December 2011.

Although the public health infrastructure of the State is weak, the public health facilities happen to be the primary providers of health care services in the State. The health care delivery system of the State follows Indian Public Health Delivery System of three-tier model. At the lowest level, Sub-Centres (SC) provides public health services. The next tier is Primary Health Centres (PHCs) where basic medical care, disease prevention and health education are provided. At the top tier, Community Health Centres (CHCs) and District Hospitals (DHs) offer specialist services. The State also has medical colleges and nursing colleges. The private sector hospitals of the State are unevenly distributed and are clustered mostly in Guwahati; and the State capital like many other underdeveloped States of India. As per the State Programme Implementation Plan (2010-11), Assam has got 131 private sector hospitals spread over 13 districts. The public health infrastructure of the State can be summarized as follows:

Table 14.2: Public Health Infrastructure

Health Facility	*Number*
Medical Colleges	4
State Level Hospital	1
District Hospitals	24
Sub-Divisional Hospitals	13
CHC	108
PHC	844
SC	4592
B.Sc Nursing College	1
General Nurse Midwife (GNM) Training Centres	15
Auxiliary Nurse Midwife (ANM) Training Centres	18

Source: SPIP 2010-11.

The Governments at the national and State level appear to be well informed of the public health status and issues of the State and have been forced to take note of the abysmal scenario existing at the grass roots level. Since 2005, under National Rural Health Mission (NRHM) a flagship programme of Government of India, efforts have been made by the State agency of this social sector initiative to improve the health of its region including Assam.

Mass Media Intervention

Mass Media are considered to be the tools for the transfer of information, concepts, and ideas to both heterogeneous and specific audiences for decades especially in the last fifty years of independent India. While in keeping with the dominant tradition of the West to use mass media as a magic multiplier of social messages, the use of mass media in the context of development especially in societies with deep rooted tradition has been a little intriguing because of the mixed results. Communicating about health through mass media has been found to be a little more complex as had already happened with family planning programme in India. Liana Winett and Lawrence Wallack (1996) argued that "using the mass media to improve public health can be like navigating a vast network of roads without any street signs—if you are not sure *where* you are going and *why*, chances are you will not reach your destination" (p. 173).

Using mass media can be counterproductive if the channels used are not audience-appropriate, or if the message being delivered is too emotional, fear arousing, or controversial. Undesirable side effects usually can be avoided through proper formative research, knowledge of the audience, experience in linking media channels to audiences, and message testing.

Sophisticated societies are dependent on mass media to deliver health information and especially electronic media have been described as "extensions of man." G.L. Kreps and B.C. Thornton (1992) believe media extend "people's ability to communicate, to speak to others far away, to hear messages, and to see images that would be unavailable without media" (p. 144).The value of health news is related to what gets reported and how it gets reported. According to Ray Moynihan and colleagues (1999).

The news media are an important source of information about health and medical therapies, and there is widespread interest in the quality of reporting. Previous studies have identified inaccurate coverage of published scientific papers, overstatement of adverse effects or risks, and evidence of sensationalism. The media can also have a positive public health role, as they did in communicating simple warnings about the connection between Reye's syndrome and the use of aspirin in children (p. 1645).

Despite the potential of news media to perform valuable health education functions, Moynihan *et al.*, (*ibid.*) conclude that media stories about medications continue to be incomplete in their coverage of benefits, risks, and costs of drugs as well as in reporting financial ties between clinical trial investigators and pharmaceutical manufacturers.

The mass media are capable of facilitating short-term, intermediate-term, and long-term effects on audiences. Short-term objectives include exposing audiences to health concepts; creating awareness and knowledge; altering outdated or incorrect knowledge; and enhancing audience recall of particular advertisements or public service announcements (PSAs), promotions, or programme names. Intermediate-term objectives include all the earlier said objectives as well as changes in attitudes, behaviours, and perceptions of social norms. Finally, long-term objectives incorporate all of the aforementioned tasks, in addition to focused restructuring of perceived social norms, and maintenance of behaviour change. Evidence of achieving these three tiers of objectives is useful in evaluating the effectiveness of mass media intervention.

Mass media, in the realm of social campaign domain are expected to perform three key functions: educating, shaping public relations, and advocating for a particular policy or point of view. As education tools, media not only impart knowledge, but can be part of larger efforts (e.g., social marketing) to promote actions having social utility. As public relations tools, media assist organizations in achieving credibility and respect among public health opinion leaders, stakeholders, and other gatekeepers. Finally, as advocacy tools, mass media assist leaders in setting a policy agenda, shaping debates about controversial issues, and gaining support for particular viewpoints.

Public Health Campaigns through Mass Media –The Conundrum

The inherent properties of various mass media have always forced the health-message designers to consider a series of questions relative to choice of channels:

- Which channels could be the most appropriate for the health problem and message,
- Are these channels the most likely to be considered as credible and accessible by audience,
- What are those channels which fit in with the objective of the programme (inform, influence attitudes, change behaviour), and
- What could be the most feasible channels considering advertisers' time and budget?

Maternal Health and Child Health: The Unresolved Quagmire

Reduction of maternal mortality, infant mortality and improvement in maternal and child health are the major goals in Millennium Development, to which India is a signatory. This goal can be achieved only if the maternal and child health improves in the States like Assam where the maternal and infant mortality is very high, which needs to be reduced.

Worldwide it has been recognized that the women in the reproductive age group (15-49 years) are the vulnerable fraction of the population due to risk connected with child bearing (WHO, 2004). Each year in Assam, approximately 680,000 women experience pregnancy and 612,000 have a live birth. Of these, an estimated maternal 2,598 deaths occur each year (PIP, 2010). The causes of maternal mortality in India are anaemia (24%), hemorrhage (23%), abortion (12%), toxemia (10%), puerperal sepsis (10%), malpositions (7%) and others (14%). This causes can be held true for Assam as well (WHO, 2004). The socio-cultural determinants also have a cumulative effect on maternal health. A correlation exists between gender/social inequity and maternal health. For example – the female literacy rate is 67.27 per cent, sex ratio is 954 females per 1000 males which exhibit a *son preference* and 36.1 per cent of the population still lives below poverty line.

Similar to national average, in Assam motherhood comes early in most of the women's lives. About 38.6 per cent of the women are married by 18 years. Median age at first birth is 20.7 per cent; however 16 per cent of the women experience motherhood before the age of 19. Women in the rural areas (17%) are more vulnerable to begin child bearing at an early age than the urban areas (13%). This shows that motherhood comes before physical maturity of the women and hence can have adverse consequences that could be related to pregnancy complications and maternal mortality.

Around the world, it has been recognized that pregnancy related care plays a vital role in pregnancy outcome. Antenatal checkup (ANC) helps in detecting and preventing pregnancy related complications. In Assam over 66 per cent of the women receive some sort of antenatal care from the health professionals but only 36 per cent receives at least three ANC. Only 38 per cent of women seek ANC during first trimester. Again, urban women are twice more likely to have three or more visits than women in rural areas.

The quality of care during ANC is poor. The common service received by the pregnant women during ANC is examination of abdomen (61%) and blood pressure (53%). The services of weight check (around 50%), blood and urine test (37 to 39%) is poor. Consumption of iron and folic acid (IFA) supplements is also poor; from 62 per cent of mothers only 16 per cent had consumed IFA for the recommended 90 days or more, only 2 per cent took de-worming during pregnancy.

During pregnancy along with ANC, there is also a need of skilled birth attendant (SBA) during and around delivery. In the State, out of all births 31 per cent takes place with the assistance from health professional and the remaining 69 per cent is delivered by trained birth attendant (TBA), relatives or untrained persons. The poor access to skilled professionals during delivery is justified by high home delivery. Around 78 per cent of the births take place at home and only 22 per cent in a health facility. Postpartum care is also a concern in Assam, only 16 per cent had postpartum check-up and 14 per cent received within two days of birth.

The situation of infant care is also grim in the State. Out of the home deliveries, many neonates develop complications just after the birth but due to not having adequate set-up at most of the health facilities (CHC, PHC), most of the neonates die at an early age. It is estimated that neonatal death contributes 65 per cent of the Infant deaths (as per study conducted by SEARCH, Gadchiroli), the same must be true for Assam too. However, the scenario of infant deaths have improved in Assam as under NRHM, a good number of initiatives have been initiated by the Government for popularizing full ANC of pregnant mothers, institutional delivery, 48 hours stay of mother at the hospital with the new born. Even to improve the sex ratio in the State, Government of Assam has initiated scheme, where the newborn girl child is gifted with Rs. 5000 (Five thousand only), which she can encash after 18 years.

A few of the schemes, which has largely contributed in improving the maternal and child health scenario in the State is discussed below.

1. Janani Suraksha Yojana (JSY)—A Conditional Cash Transfer to Promote Institutional Delivery

In 2005, Government of India launched Janani Suraksha Yojana (JSY) to incentivize women to give birth in a health facility (Lancet, 2010). The State of Assam has also implemented this programme to increase births in health facilities and is not limited only to the public health facilities. According to the JSY guideline, the State has also included empanelled/accredited private hospitals under PPP; the women giving birth in these health facilities are entitled for the incentive. All women in the State irrespective of socio-economic status and parity are eligible for the cash benefit (MOHFW, 2006), as the State belongs to Low Performing State (LPS). The women in rural areas receive incentive of Rs. 1,400 and Rs. 1000 in rural areas (MOHFW, 2006). This lone initiative has helped State in multiplying institutional deliveries to a large extent. The institutional deliveries, which were just 1,49,003 in the year 2005-06 went to more than 4,53,356 in the year 2011-12 and this quantum of jump in the institutional delivery has saved lives of many mothers and infants.

1. *Majoni—A Special State Assistance Scheme for New-born Girl Children*

Title	*MAJONI—Special care for the girl child*
Geographic area of Coverage	In entire Assam
Target Population	Girl Child
Eligibility criteria	Up to two child norms
Project Life	Launched on17 February, 2009
Budget	Rs 100 crore per Year
Implementing Partners	Assam Bikash Yojana (ABY)

Objectives: The key objective is to provide special attention to girl child. In Assamese, beloved girl child is named as "**Majoni**".

Many a time, because of poverty, girl child is considered as burden and parent want to marry her as soon as they can. **Majoni** also aims to usher in radical changes in caring the girl child in the State through eliminating all kinds of gender bias and help the girl child grow with dignity.

Strategies: The scheme covers only the girl child born at a government hospital. On having institutional delivery, the girl (up to two child norms) gets a fixed deposit of Rs. 5000 (five thousand) in her name, which can only be encashed once the child completes 18 years of age. The photograph of both parents will have to be submitted at the time of issue of the fixed deposit and thumb impression of the child. A certificate is issued by the bank while renewing the Fixed Deposit in the name of the Girl Child wherein her photograph at 10 years, i.e., time of renewal of Fixed Deposit (photograph attested by any executive Magistrate or Revenue Officer) and a 'life certificate' will be stamped by the bank. Photograph at the time of maturity to be attested by the local Revenue Circle Officer/Executive Magistrate. In keeping with the policy of the government, the child beneficiary should not be married before the age of 18. However a girl child born to a mother staying in a paying cabin or a private hospital is not eligible for the assistance. The scheme has come into effect in Assam from 1st February, 2009. The beneficiaries/nominee can encash the matured policy at any branch of the scheduled bank on identification by Executive Magistrate or local revenue officer. In the event there is no claim after 18 years the maturity value against the Fixed Deposit will revert back to the Assam Bikash Yojana (ABY) account of the issuing bank.

2. *Mamoni-Nutritional Food to Pregnant Women*

Title	*Mamoni—Nutritional Food to Pregnant Women*
Geographical area of coverage	In entire Assam
Target Population	Pregnant Women
Project Life	Started on 1st March, 2009
Budget	Rs. 80 crore per Year
Implementing Partners	Assam Bikash Yojana (ABY)

Objectives: Recognizing the fact that "***a healthy mother begets***

a healthy child", Government of Assam under Assam Bikash Yojana (**ABY**) launched the innovative scheme namely **Ma-moni**, which allows every expectant mother to get Rs. 1,000 for supplementary nutrition during pregnancy. To ensure healthy pregnancy, the importance of balanced diet along with regular antenatal check-up is important, especially calcium and iron rich diet is essential for the growth of a baby in the mother's womb.

Strategies: The scheme can be availed by the mother up to two children. The expectant mother will receive a copy of the "**Ma-moni**" booklet at the first antenatal check-up, immediately after registration. During her second ANC, the expectant mother will also receive an A/C payee cheque of Rs. 500 and a voucher for free transfer to the hospital. The expectant mother will receive Rs. 500 at the time of the third antenatal check-up. For second and third ANC, mother will have to carry MCH card. The cash assistance can be availed of during second and third ANC in the government health institutions, which can be Sub-Centre including village health and nutrition day (VH & ND), PHC, CHC, District Hospital. The staffs of the health institutions wherever mother has registered herself will be responsible to provide information about the scheme and its importance. The pregnant woman will be entitled to receive the cash assistance only if she belongs to that village, i.e., her husband's village. Simultaneously, she will receive the remaining amount only if she completes her third ANC in the same health institution where she did her first and second ANC. In cases, where a pregnant woman is planning her delivery in the health institution other than the health institution in her husband's village/block/district (e.g., her maternal village) and accordingly has planned her third trimester ANC in other village/block/district. In those circumstances, it is prerequisite that she has registered herself in the SC/health institution of her husband's village and has received Rs. 500 along with ANC materials. The ANM/doctor of the SC/health institution will certify that she can avail the assistance amount for third trimester from other health institution but the name of the block/ district should be specified. The SC/health institution will disburse the cheque of remaining Rs. 500, if the pregnant woman carry the certificate and the MCH Card along with her. The ANM will prepare a village-wise list of pregnant women registered and hand it over to the ASHA of the respective village for follow-up. ASHA in the village will be responsible for follow-up of the pregnant women and will ensure through counselling that the amount is utilized for the given purpose.

Service Utilization–Institutional Delivery

Assam has shown considerable progress between 2005 and 2012, institutional deliveries have been gradually increasing from 23 per cent (NFHS 3) to 64.4 per cent (CES). This positive trend depicts that the State has been able to improve access of pregnant women to health care.

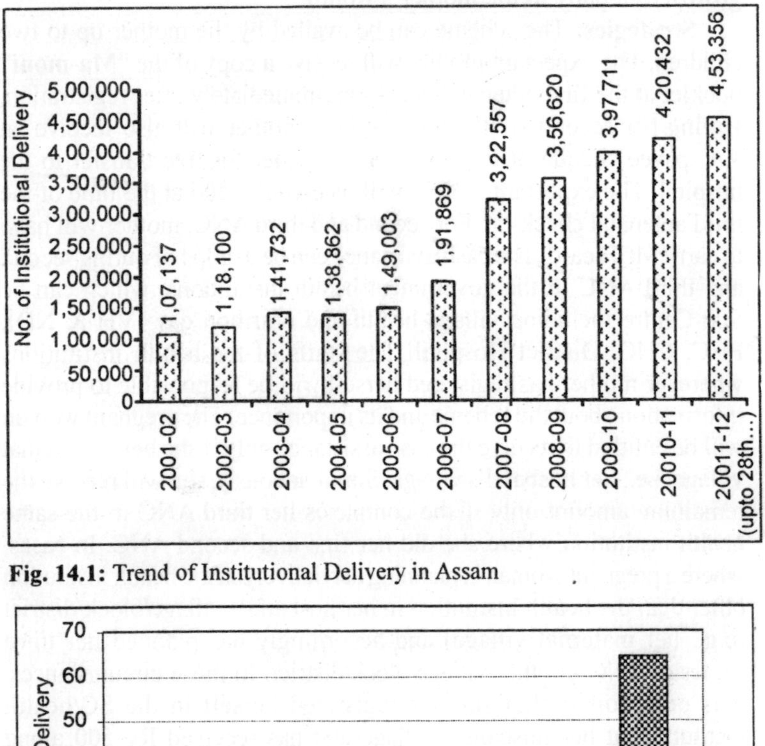

Fig. 14.1: Trend of Institutional Delivery in Assam

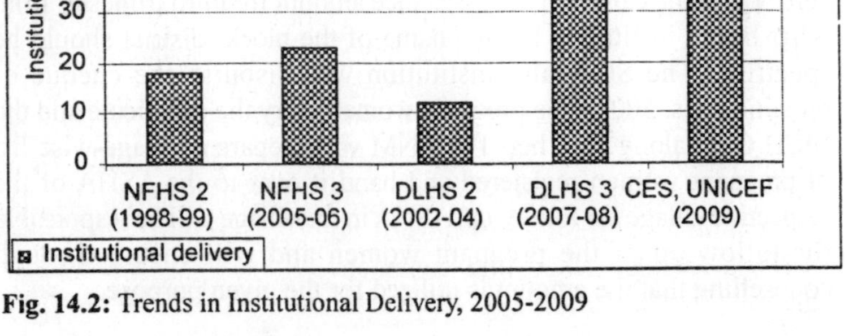

Fig. 14.2: Trends in Institutional Delivery, 2005-2009

However, the increase in the institutional delivery still maintains the urban and rural disparity in service utilization. The figure shows that gap between urban and rural have been reduced since 2005, still a gap of 20 per cent remains between the two areas. This describes that a focused intervention concentrating on rural areas is the need of the hour.

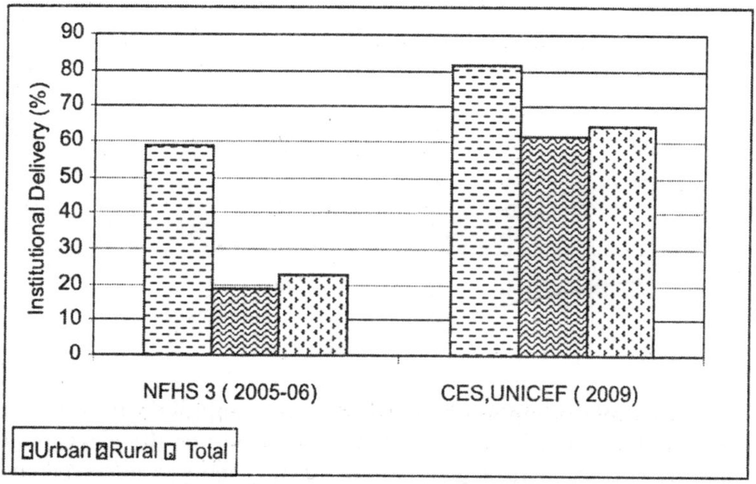

Fig. 14.3: Urban and Rural Comparison of Institutional Delivery, 2005-2009

Moreover, while the delivered women are discharged in less than 48 hours without admitting them for post-delivery (JRM 6, 2009), 28 per cent of the women are discharged in less than 24 hours and 26 per cent within 1 day (CES, 2009). This highlights that although the institutional delivery has increased but this increase in service utilization is not complemented by post-natal care. There is no provision or initiative on post-natal practices at the facilities (JRM 6, 2009).

All these programmes/initiatives would not have yielded any result if mass media would not have generated awareness among community to avail of all these benefits The advertisement on these new programmes come to different newspapers through which people came to know about these programmes and started availing these benefits. The role of ASHA (Accredited Social Health Activist) in generating awareness also needs to be specially mentioned. These ASHAs are also taken into confidence for propagating any new messages among the masses. The ASHAs through interpersonal/group communication take the messages to the target group and make them convinced to take the benefits of the programmes. This mass media intervention has shown result not only in Assam but also in entire North-Eastern States, which is seen from Table 14.3 given below in the context of IMR.

States/ Country	*Dec'11*	*Jan'11*	*Oct'09*	*Oct'08*	*Oct'07*	*Oct'06*	*April'06*
India	47	50	53	55	57	58	58
Arunachal Pradesh	31	32	32	37	40	37	38
Assam	58	61	64	66	67	68	66
Manipur	14	16	14	12	11	13	14
Meghalaya	55	59	58	56	53	49	54
Mizoram	47	36	37	23	25	20	19
Nagaland	23	26	26	21	20	18	17
Sikkim	30	34	33	34	33	30	32
Tripura	27	31	34	39	36	31	32

Source: SRS, Dec - 2011

Epilogue

Analyzing all the important footsteps made in improving the maternal and child health status, it can fairly be argued that the mass media appeared to have played an important role in making all these schemes popular and generated an awareness among masses to encourage them to avail the benefits of all these schemes and thus improving the maternal and child health status in the State. The situation demanded a proper choice of mass media and also a proactive role of mass media, and the combined effort of these two helped messages related to all these schemes reach the remote villages generating demand of the community to have the benefits of all these schemes. The concerned agencies, virtually forced by the situation had no other choice but to respond to the community demand and to extend the services highlighted under the scheme to the beneficiaries. The reduction in the maternal and infant death in the State to a large extent is the classic example of mass media intervention in this regard. The MMR, which was 490 as per SRS 2004-06 has come down to 381 as per Annual Health Survey, 2010-11. An independent evaluation, conducted by Regional Resource Centre for North-Eastern States (RRC-NE) in 2010-11 found out that the State's MMR is 333. Although, it has come down, but it has to be further brought down to achieve the NRHM goal. The Infant mortality rate has also reduced from 66 (SRS, April '06) to 58 (SRS, Dec '11), which is seen as a big achievement. The situation of infant mortality in other North-Eastern States have also changed, which is shown in the above table. The role of mass media in tackling the challenges posed by underdeveloped health sector in the State has been a laudable one.

However, thereby hangs a tale, which needs to be told before one becomes too optimistic about the use of mass media and it is so called guaranteed success in health sector. When one gets engaged in the process of evaluating the capability of mass media intervention in the health sector, one must also understand the grey areas which demand a certain amount of clinical inspection. Mass media, although, tom tomed as a magic multiplier has been rejected in many cases as it failed to caste a spell on the audience for their sheer inability to access or afford it due to both financial and literacy constraint. The idea of information rich and information poor are the ideas that germinated in mass media. While we have reached the era of new media, the burdens of poverty and illiteracy are as yet strenuous enough to stop the dispersal of any message. There have been efforts all over the world, especially the third world countries to use a right mix of mass and alternative channels of communication. Problems with mass media essentially are twofold; it treats a heterogeneous audience as a homogeneous one as in the case of advertising and including social campaigns converting the social initiatives into a case of social marketing, a major factor responsible for rejection of mass media as dominant wheels of development. Secondly, it uses a uniform code for the dissemination of messages which fails to transcend the barriers imposed by the cultural ethos of a tradition bound society. The development planners in the health sector in India on the whole and in particular in the North-East must be able to put too many eggs into the basket of mass media. It should not be treated as a panacea or the golden boy of social campaign. The campaign strategy in the health sector must be homogenized and Indianized. It has to be even so in the North-East, a place full of multicultural ethos. Multiculturalism by its very propensity rejects any effort of homogenization and the dismal scenario of any social campaign with a blanket strategy is bound to fail in this region.

While designing a message for local populace, the health development planners have to seek the help of health service seekers. It might be a little too much to expect from the development planners to reverse the well defined top-down approach, but it has to be accepted in a region which has a tradition to live in terms of closeted societies. The messages, without a shade of doubt, has to be culture specific especially in a region where too many cultures are vying for attention and in fact is dominated by the politics of identity.

To put it more succinctly, unless and until, alternative communication channels like community radio or folk formatised messages are used hugely; it is difficult to reach the hitherto unreached. Even though, the success of mass media in disseminating developmental messages in the health sector of Assam is remarkable compared to the abundant stories of failures in different sectors and different parts of the country of the same communication tool, the fact remains that conventional media is being increasingly treated as a dead horse in the domain of social communication without fail and without exception across the world when communication in the age of Dispersed Media is becoming increasingly culture specific.

REFERENCES

Amarsinghe (2001). Mobilizing Rural Women for Development: Experience of Social Mobilization Programme in the Hamabantota District, Sri Lanka.

Askwer, I. (1986). *The Community Participation Approach in Family Planning Programmes*. London: IFFP.

Barua, Jayanti (2001). *Social Mobilization and Modern Society*. New Delhi: Mittal Publications.

Bray, Mark. (1996). *Decentralization of Education: Community Financing*. Washington, DC: World Bank.

Cohen, Sylvie I. (1969). Mobilizing Communities for Participation and Empowerment, in S. Cohen (Eds.) *Participatory Communication for Social Change*. New Delhi: Sage Publication.

Government of Assam (2005). *State Programme Implementation Plan (SPIP) of NRHM, Assam for the Year 2005-2012*. Guwahati: Government of Assam.

Government of India (2005). *NRHM Framework and Implementation*. New Delhi: Ministry of Health and Family Welfare.

John, S. (2003). *A Guide to Empirical Research in Communication*. New Delhi: Sage Publication.

Kreps, G.L., and Thornton, B.C. (1992). *Health Communication Theory and Practice*. Prospect Heights, IL: Waveland Press.

Moynihan, R. (1999). Coverage by the News Media of the Benefits and Risks of Medication. *New England Journal of Medicine*, Vol. 11.

Pattanaik, B.K. and Singh, Kulwinder. (2002). Community Participation, Panchayati Raj and Rural Health Care. *Kurukshetra*, Vol. 46 (6).

Prasad, R. R. (2003). Social Mobilization: Concept, Meaning and Approach. *Kukukshetra,* Vol 45(2).

Roy, G. L. (1974). Extension Communication in India., Kolkata: Atma Ram & Sons.

Winett, L. B., and Wallack, L. (1996). Advancing Public Health Goals through the Mass Media. *Journal of Health Communication*. Vol. 51 (9).

15

Understanding the Sexual and Reproductive Health Needs of Pregnant Women Living with HIV/AIDS: Some Reflections from Manipur

Grace Hmar

Introduction

There is a growing feminization of the HIV/AIDS epidemic in India with 38.4 per cent of those living with the virus being women (Willson and Cleason, 2010). HIV affects or potentially affects all the dimensions of women's sexual and reproductive health-pregnancy, childbirth, breastfeeding, abortion, use of contraception, exposure to, diagnosis and treatment of Sexuality Transmitted Infections (STIs) and their exposure to sexual violence. Infection with HIV also affects the sexual health and well-being of women.

More and more of them are accessing antiretroviral treatment, extending their lives and their productivity. An important part of their lives, as for any human being, is sexuality and reproduction. Like everyone else, they have a right to a satisfying, safe, healthy sexuality and reproductive health. However, research suggest that most HIV positive women do not receive the information, support, referrals and related services necessary to meet their sexual and reproductive health needs as HIV positive women (Koegh, R.; Allen, S., and Temahagili, B., 1994). Instead, HIV positive women experience stigma and discrimination related to their sexuality and reproductive choices.

Women living with HIV have always wanted and deserved to have healthy, satisfying sex lives, to bear and raise children and to protect themselves and their partners from unwanted pregnancies and sexually transmitted infection (STI) including HIV.

Sexual health is defined as a state of physical, emotional, mental and social well-being in matter related to sexuality. It is not merely the absence of disease, dysfunction or infirmity. It is an integral aspect of human development and maturation throughout the life cycle. Complex factors influence sexual behaviour. These factors effect whether women's expression and experience of sexuality leads to sexual health and well-being or places them at risk of ill-health. Unfortunately, rather than women having satisfying and safe sexual experiences, their sexuality is often the cause of distress and characterized by unsafe or harmful sexual practices that lead to adverse health outcomes.

The term "reproductive health" was first used in international policy agreement at the United Nations International Conference on Population and Development, held in Cairo in 1994. The conference's Programme of Action defined the terms as follows: A state of complete physical, mental and social well-being ... in all matters relating to the reproductive system. Reproductive health, therefore, implies that people are able to have a satisfying and save sex life and that they have the capability to reproduce and the freedom to decide if, when and how often to do so.

Policy debates have largely ignored the sexuality of women living with HIV/AIDS and programmes to the extent that they addressed sex and reproduction at all, were generally limited to helping pregnant women avoid transmitting the viruses to their children. Given that most HIV transmission occurs through sexual intercourse, it is critical to include a sexual and reproductive health lens in HIV programming. Most programmes currently revolve around access to antiretroviral and other AIDS-related treatment, and hospital and home-based care for those with HIV and AIDS-related illness. In relation to sexual and reproductive health care, HIV prevention predominates.

The sexual health of women is an important component of Sexual Reproductive Health services and is often overlooked, particularly for women living with HIV. Discussions of Sexual Reproductive Health Services for women living with HIV often revolve around controlling fertility and ignore HIV-positive women's needs for services that include attention to safe and healthy sexuality and a desire for children. Because women living with HIV are more vulnerable to rights abuses, for example forced contraception or coerced sterilization, ensuring that their sexual and reproductive health needs are met is critical (Wilcher and Cates, 2009).

Several factors, however, have moved the international community to consider ways to meet broader needs. Sexual and reproductive health in general has received increased global attention in the years following the 1994 International Conference on Population and Development (the Cairo Conference). Many advocates have pointed out that improved sexual and reproductive health are essential to meeting the Millennium Development Goals agreed to by world leaders in 2001. At the 2005 world summit of the United Nations General Assembly, United Nations member-states committed to achieve universal access to reproductive health by 2015 as a means to reaching the MDGs. In January 2006, at the UN Commission on the status of women, a joint statement was submitted regarding the need for ensuring the sexual and reproductive health and rights of women affected by HIV. It was an essential component of the enabling environment for achieving gender equality and the advancement of women.

Review of Literature

Policies regarding pregnancy and childbirth in the light of HIV have been prescriptive, emphasising reproductive control and prevention. Due to considerations of women's health and the risk of vertical transmission to the foetus, HIV-positive women are advised to avoid becoming pregnant, or to terminate their pregnancy (WHO/UNAIDS, 2003). This policy has been noted to pose substantial challenges for women, as the ability to have children plays a vital role for very many women's social status. Ekra *et al.*, (1999) and Tabi and Frimpong (2003) indicate that childbearing may indeed be more important to women of many cultures than a fear of illness or death. Therefore, despite knowledge of being infected, many women may become pregnant or continue their pregnancy.

AIDS Analysis Africa (1998) and Sakyi (1999) write that in Malawi and in most other sub-Saharan African countries, a woman's social status is directly linked to her ability to produce children. Children epitomise meaning of a woman's life. A woman without children looks upon herself as incomplete, as useless. In a married couple, the responsibility for infertility is viewed as the female partner's problem, and the community may sometimes subject barren women to ridicule. The promotion of condoms to prevent the spread of sexually transmitted diseases such as HIV and STI's will by most women and men simultaneously be perceived as being in direct opposition to the fulfilment of a powerful social norm (Mill, & Anarfi, 2002).

Maposhere *et al.*, (2002) have carried out a study of particular relevance for the present study. They explored HIV-positive women's views and experiences of pregnancy and childbirth through a survey among 209 HIV positive women in Zimbabwe. Sixty-four per cent of the 209 women reported that the community and relatives did not expect HIV-positive women to bear children and 89 per cent of these women (the 64%) said that HIV-positive women should not have children. Almost all the women who expressed these views however were over 25 years of age and had already had at least one child. The reasons given for not wanting more children were that they should give priority to the children they already had, that they were worried about the economic impact another child would have on their household, and the extra burden it would pose on their time and energy. They were also concerned that pregnancy and childbirth would affect their health, and that more children would be left orphaned. However, HIV-positive women who did not have any children, or those who were young, in a new relationship, or had a husband who had paid bride price ("*lobola*"), indicated a strong desire to become mothers, and chose to become pregnant despite knowledge of their HIV-positive status. This was so even in the face of previous experiences of miscarriages or stillbirth (Feldman *et al.*, 2002).

Meena (1992) indicates that this enormous value placed on fertility may have an impact on the decision-making by HIV-positive women during pregnancy. In the sense that an infected woman may decide to proceed with her pregnancy despite knowing the risks to her own health and the risks of transmission to her baby, in order to fulfil her own desire as well as the societal expectation to bear children.

Gilks (1998) and De Bruyn (2002) have indicated that HIV-positive women who decide to carry their pregnancy to term also have reported adverse consequences. Apart from the additional physical hardships that a pregnancy brings, HIV-positive pregnant women do often, as we have noted above, face hostile responses from health-care workers, partners and family. HIV-positive women are told not to be sexually active. They may also have difficulty in gaining access to hospitals for delivery, and sometimes experience differences in quality of care.

Many HIV-positive women report healthcare provider hostility and judgmental attitudes regarding desires for family planning (POLICY Project, 2006). Healthcare providers often believe that HIV-positive women should not be sexually active (Kyomuhendo and Kiwanuka, 2008). When Family Planning and Reproductive Health (FP/RH) services are offered, providers frequently limit the number of options for HIV-positive women (Kyomuhendo and Kiwanuka, 2008). This may be due to a lack of clear information and guidelines on which services are appropriate for HIV-positive women or a manifestation of stigma and discrimination. Providers themselves are part of broader communities where there may be societal stigma and discrimination toward people living with HIV (PLHIV), especially toward sexually active HIV-positive women who are considering whether to have, delayed, or not have children.

Studies in Zimbabwe and Thailand show that HIV-infected women may not be able to terminate pregnancies because they lack information and access to safe, legal, and affordable abortion services (de Bruyn, 2003). On the other hand, HIV-infected women often are counselled to terminate a pregnancy or to avoid having children, a practice that compromises their reproductive rights (de Bruyn, 2002). In addition, a woman who decides she wants to conceive "should be offered advice about how to do so while minimizing the risk of infecting her partner or of becoming reinfected by him, by learning to recognize when she is most likely to conceive and avoiding intercourse or using barrier contraception at other times" (Bury, 1989).

Method

Sample

A total of 250 women living with HIV participated in the study. 115 were pregnant at the time of interview and 135 had already delivered. A qualitative design using in-depth interview and focus group discussions was used to explore the experiences of HIV positive pregnant women in regard to their sexuality and reproductive health needs. Common themes and patterns emerged from women's narrative.

Study Setting

The study was conducted in Churachandpur district, one of the high HIV prevalence rate among pregnant women.

Data Collection and Analysis

The women were contacted through the staff of the NGOs they were associated with and consent were taken from the women living with HIV. A non-directive approach was used and women living with HIV were encouraged to give detailed responses with minimal intervention from the interviewer side throughout the interview, a monologue was encouraged with minimal verbal and non-verbal probes. All the in-depth interview and focus group discussions were conducted in native languages and than translated into English. During transcription, all personal identifiers were removed and a subject/interview code was assigned to protect confidentiality. The data collection and analysis took place concurrently. This allowed employing transcribes information for subsequent data collection and determine data gaps and the need for more interview sessions. The data was clean, coded and analysed.

Findings

1. Uncertainties related to prevent the Future Pregnancies: Women living with HIV want to avoid pregnancy for a variety of reasons: Fear of passing HIV infection and feelings of responsibility, do not want to take risk, fear of being judged by others (especially health care providers), already has children, and economic constraints. However knowledge and actual use of contraceptives is very limited among women infected with HIV. Figure 15.1 indicates that most of the women reported that they surely did not want to become pregnant in the future. Most of the women were not sure of how to go about not having more children. Some of the women whose both the partners are positive highlighted that they were not 'frequently' using condoms. They thought they personally did not need to use condoms, and that use of condom would imply both a physical and emotional barrier in the relationship with their partner. It would lead to conflicts in the family once the issue is introduced. When asked in the following way; "You have told me that you plan not to have any more children. Can you please tell me the action you will take to prevent future pregnancies?" answers such as the following came:

"My future plans are that I should never become pregnant again, but I am not sure about the family planning methods. I have to discuss with my husband."

—HIV post-partum woman.

"We have not yet decided. I do not know much about family planning methods."

—HIV post-partum woman.

"Yes, I do not want to have any more children, but I do not know the method I can use"

—HIV post-partum woman.

"As I told you, I have problems because my husband is forcing me to have children or he will abandon me, so I cannot use any kind of method. I will see once the child is born..."

—HIV post-partum woman.

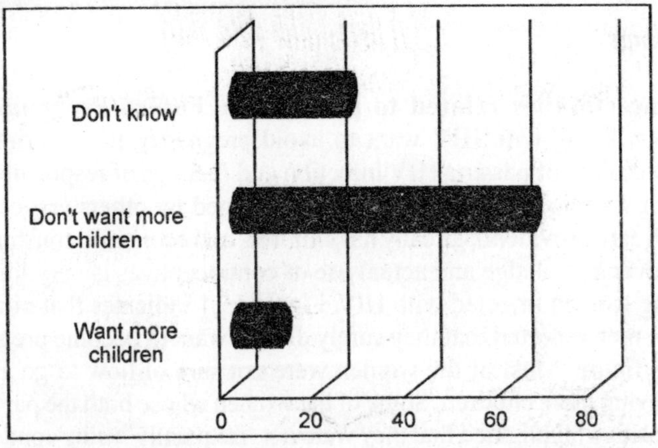

Fig. 15.1: Fertility Desires of HIV-Positive Pregnant and Post-partum Women

For most of the women who did not want another child, the main reason was concern about the child's future without them. According to one HIV-positive post-partum woman, "I don't want more children, if something happens to me who will look after my children". However, for some of the women, even if they did not want to have more children their husbands felt otherwise and pressured them to have more children.

> *"...even now [after knowing my HIV status] he wants a child. He says now it is nearly one year we shall have one more child. He is very fond of children. It is I who is trying to avoid, I know it is difficult to bring up the child what for men they will not know what we undergo. Knowing the child is positive still he wants a child and I already have 3 children."*
> —HIV-positive post-partum woman

> *"I am going to die so why do I want more children. This child I am having is because my family [husband and mother] forced me to".*
> —HIV-positive pregnant woman

In some cases it appears that the health provider advice in favour of sterilization may be influenced by their opinion or views about the child's future as a potential orphan.

> *"They [health care provider] said to have an operation and not to have another baby as we would leave them as orphans in this world—who will take care of them?"*
> —HIV-positive post-partum woman

Some of the women in the qualitative interviews mentioned using natural methods of pregnancy prevention, but this was not a major method that emerged from the quantitative data.

> *"....when we join [sexual intercourse] there is a liquid [semen] that comes out of the man....at the end time he will leave the liquid [semen] out. Because of this the baby will not form. We are doing this now".*
> —HIV-positive post-partum woman

2. Pregnancy and Child Birth: Giving birth and motherhood play a very significant role in the social status and self identity of women in many cultures. Child bearing may indeed be more important to women of many cultures than fear of illness or death (Tabi and Frimpony, 2003). Therefore, despite the knowledge of being infected, many women may become pregnant or continue their pregnancy, in order to fulfil her own desire as well as the societal expectation to bear children (Meena, 1992). Various reasons were cited by the women for wanting to have baby:

(A) *Sustaining the Family Genes*: Some women talk about the need to have a baby from the perspective of sustaining their family.
"My parents always wanted me to sustain our generation. I would follow that".
—HIV positive pregnant woman
"We want to leave an heir behind us...it is natural isn't it?"
—HIV positive pregnant woman.

(B) *Need to Experience Motherhood*: A woman asked, "What is their in life in which there is no baby?". Another woman said, "Whether negative or positive, a desire to have a baby is present in all women". This perspective strongly reflect societal norms that portrait motherhood and child bearing as one of the essential roles for woman.

(C) *Societal Influences*: The fear of being labelled as "barren" influence decision making to have a baby even if a women personal choice might be not to have a baby. As a woman explained... "It has been since two years we got married, found to be positive in the first year itself, my mother-in-law keep asking me when I am going to give baby". Thus in the absence of disclosure of HIV status to their family members and not wanting to be labelled as barren, some HIV positive women feel compelled to fulfil the expectations of the family members.

(D) *Need to have a Boy Child*: Some of the women express to have another child, hoping that it would be a boy. Another HIV positive woman expresses her dilemma regarding having a baby. "We have a girl child but my husband wants a baby boy. So he will not listen to what I say".

However, positive women faced judgmental attitude from health care workers who believe that they should not have children, either because of the risk that HIV will be passed on to their baby during childbirth or breastfeeding or out of concern for the child if its parents die prematurely from AIDS. Women who do desire children are thus deemed irresponsible or deviant. The result is often that women are put under intense pressure from health workers to have an abortion or sterilized.

3. Contraceptive and Safer Sex: Health care services must engage with the reality that women living with HIV continue to have sexual relationship either through desire or necessity. This presents certain challenges. Most of the currently pregnant women mentioned that their Prevention of Parents to Child Transmission (PPTCT) staff advised them to use condoms during sexual intercourse with their partners. However, only about two-thirds of these survey respondents reported that they and their partners were using condoms during sexual intercourse. The reasons for not using a condom varied and included "no need to use with husband" and "husband does not like it."

Some of the women also explain that "no option for us". No discussion on any contraceptive other than condoms. They don't go to the next level.

"I have never used [a condom] with my husband. We never felt the need to use and are not using now also".

—HIV-positive pregnant woman

"My husband does not like using condoms so we do not use them".

—HIV-positive pregnant woman

In qualitative interviews with pregnant and post-partum women, some mentioned that they were not always able to use condoms with their partners even though their service provider had counselled them so.

"We do use condoms, but not alwcys. We know it is best to use condoms but sometimes we forget and don't use it."

—HIV-positive pregnant woman

"...get condoms from the NGO and we used it quite often. But

sometimes when my husband is drunk, I am not able to negotiate with my husband. And moreover I don't want to create chaos in the house fighting with him so I had to oblige."

—HIV-positive pregnant woman

"I never used condoms with my husband. With my sexual clients I have used it sometimes because they only wanted me to use condoms. Otherwise I never made an issue out of it. I know the NGO staff told us to use condoms always. But it is not always practical all the time."

—HIV-positive post-partum woman and current drug user

"From the time I have been declared positive we have been using condoms. Because my husband is negative I don't want him to be infected".

—HIV-positive post-partum woman

"We are using condoms for two reasons. One is not to have more children and the other is not to get my husband infected. I want him to live long."

—HIV-positive post-partum women

4. Access to Safe Abortion: Even where contraceptive services are available unintended pregnancies still happen for a variety of reasons. Contraceptive may fail, may not use contraceptive for fear of side effect or sex may be coerced or forced. Many women will seek to terminate these pregnancies. However, many HIV infected women may not be able to terminate pregnancies because they lack information and access to safe, legal and affordable services (de Bruyn, 2005).

"I was tested positive during antenatal check up. I wanted to have an abortion as I already had two children but I cannot afford the expense charge by the doctor for abortion".

—HIV positive pregnant woman

"I wanted to have an abortion but what will other say? They will not understand, they will say she wanted a baby, she had, and now she does not want. I do not want to face that situation. I have not disclosed my status to any one".

—HIV positive pregnant woman

Thus fear of other reaction and fear around disclosure of HIV positive status may lead some women to have a child even if they might prefer not to continue the pregnancy.

"As advised by my PPTCT counsellor, I went for abortion at the family welfare services, the staff there refused to abort my child after knowing my status. They advised me to go to private clinic. How can a person like me afford to go for abortion at private clinic?"

—HIV positive pregnant woman.

Discussion and Conclusion

High-quality programmes and services that positively address sexuality and promote the sexual health of women living with HIV/AIDS are essential for women living with HIV/AIDS to have responsible, safe and satisfying sexual lives, especially in places severely affected by HIV. When the women actually wish to limit the number of children, to prevent becoming pregnant, and to employ contraception, the findings indicate that they had little say. Their power in negotiation and decision making regarding reproductive choices was extremely limited. It is crucial that HIV positive women are supported to make their own reproductive choices about whether and/or when to have children, free from discrimination, coercion and violence. To support HIV positive women in considering their reproductive choices, accurate, impartial and accessible information is required. Women living with HIV should have greater access to female condom and female oriented method of prevention such as mircobicides in order to have control over prevention intervention without totally dependent on men. It recognises that the male condom is a male control technology and it take in to account the imbalance in power in sexual interaction that makes it difficult for women to negotiate condom use by providing women with an alternate women initiated technology. Particular attention need to be given to ensuring that safe and affordable abortion is available and accessible to the full extent allowed by law to women living with HIV who do not want to carry a pregnancy to term is essential to preserving their reproductive health.

Findings also indicate that sexual and reproductive health and HIV services are often provided through separate, parallel structure. One way to ensure that women living with HIV have adequate to sexual reproductive health care is to link or integrate services. This could mean that HIV services and sexual and reproductive health at single site or simple that health workers have the knowledge and skills to provide an appropriate basic package of services and refer patients for other necessary care that is not provided at the site.

The attitude of the health care providers in terms of discrimination experience by the women is inexplicable. Health care staff requires the skills, knowledge and understanding to treat positive women without being judgmental and to provide them with the information they need to make informed choices. It is also essential to assess and acknowledge fear and risk of health care staff, and then develop work place policies that ensure staff safety and respect health care workers rights. It is essential that prevention goals do not outweigh the rights of women as patients. Women will be more inclined to participate in programme if they feel that their choices will be respected.

REFERENCES

de Bruyn, M. (1992). Women and AIDS in developing countries. *Social Science and Medicine,* 34(3), 249-262.

de Bruyn M. (2002). Reproductive choice and women living with HIV/AIDS. Chapel Hill, NC: Ipas, Retrieved on February 2011 from http://www.ipas.org/english/publications/international_health_policies.asp.

de Bruyn M (2005). *Reproductive rights for women affected by HIV/ AIDS? A project to monitor Millennium Development Goals 5 and 6.* Retrieved on October 2011 from http://www.ipas.org/Publications/Reproductive_rights_for_women_affected_by_HIV-AIDS_A_project_to_monitor_Millennium_Development_Goals_5_6.aspx.

Feldman, R.; Manchester, J.; and Maposhere, C. (2002). *Positive Women: Voices and Choices—Zimbabwe Report.* International Community of Women living with HIV/AIDS.

Gilks, C.; Floyd, K.; Haran, D.; Kemp, J.; Squire, B.; and Wilkinson, D. (1998). *Health and Population Occasional Paper Sexual Health and Health Care: Care and Support for People with HIV/AIDS in Resource Poor Settings.* Liverpool: United Kingdom.

Koegh, R.; Allen, S.; and Temahagili, B. (1994). The social impact of HIV infection on women in Kigali, Rwanda: A prospective study. *Social Science and Medicine,* 38, 1047-1053

Meena, R. (1992). *Gender in Southern Africa. Conceptual and Theoretical Issues.* Harare: Sapes Books.

Mill, J. and Anarfi, J. (2002). HIV risk environment for Ghanaian women:challenges to prevention. *Social Science & Medicine,* 54, 325 –337.

Sakyi-Oheneba, Y. (1999). *Female Autonomy, Family Decision Making, and Demographic Behaviour in Africa.* Wales: The Edwin Mellen Press.

Tabi, M & Frimpong, S. (2003). HIV-infection of women in African countries. *International Nursing Review,* 50, 242-250.

UNAIDS. (2003). Report on the Global AIDS Epidemics. Geneva: UNAIDS

Wilson David and Claeason Mariam (2010). "Dynamics of the HIV Epidemic in South Asia". In Markus Haacker and Mariam Claeason (eds.), HIV/ AIDS in South Asia: An Economic Development Risk, Jaipur, Rawat Publications.

WHO (2003). *Safe Abortion: Technical and Policy Guidance for Health Systems.* Geneva: WHO

Wilcher Rose and Willard Cates (2009), "Reproductive Choices for Women with HIV" *Bulletin of the World Health Organization*; 87:833-839. Retrieved on 8 October 2011 from http://www.who.int/bulletin/volumes/ 87/11/08-059360/en/

16

Women and Development Activities: A Sociological Study in Karbi Anglong District of Assam

—*Jyoti Prasad Saikia & Anannya Gogoi*

"Without the progress in the situation of women, there can be no true social development. Human rights are not worthy of the name if they exclude the half of humanity. The struggle for women's equality is part of the struggle for the better world for all human beings and all societies"

—*Bautras Bautras Ghali*

Introduction

The development of a society generally depends upon proper utilization and quantity of its human resources. Therefore, there is a need to convert human beings into human resources. The term "human beings" include both men and women, since inception the functions of men and women have been separated in the society. The division of labour on the basis of sex is evident everywhere throughout the country. But both men and women have been contributing a lot for the development of society. The term 'development' consists of two common concepts, i.e., bipolar and focal. As per understanding of socio-economic perspective, bipolar concept, i.e., indicates life chances and life style of human beings. On the other hand "focal concept" includes technological aspect. In this study, the developmental activities mean by any kind of activities which are used as means to improve their economic condition as well as their family. Considering this as the economy of Karbi society depend on agriculture, it deeply observed that Karbi women engage themselves in agricultural activities to improve their economic condition, so women cannot be undermined from the point of economic development. Statistics also shows that women are the backbone of the agricultural sector, amounting 70 per cent of the agricultural labour responsible for 60 per cent agricultural productions. Thus recognition of the role played by the women in agricultural sector is fundamental to rural development and more significantly crucial for the development of women and fulfilment of their economic potentiality. Again the Karbi women are busiest group of people throughout the North-East region of India. They work hard from "dawn to dusk", if not more than at least at an equal level with the male members of the family. On the other hand if it is noticed that the women in Karbi society play dominant part in managing entire household affairs. Women have got paramount importance in deciding the day-to-day affairs of each and every Karbi family. This fact is significant as Karbi tribe is one of the many matrilineal tribes of North-East India. The programme of most effective activities in everyday working days has been formulated by the women. Interestingly the women only do preliminary rough work and thereafter methods of implementation of works are designed and formulated by the women. Women remain present in the work field and take active part simultaneously with the men. Generally in the Karbi society women have been involving in the developmental activities due to rigid implementation of customary practices. But though the majority of Karbi women are engaged in subsistence agricultural activities yet their status has not improved and they still remain as a disadvantaged group. It is worth mentioning that women were confined with some particular activities. In Karbi society it is observed that women involve in fencing, plantation, weeding, irrigation, harvesting and to some extent soil cutting. Therefore, involvement of Karbi women in agricultural operation means the involvement in developmental activities. The agricultural products are only the source of their economy. They engage in agriculture as prime income generating activities. It helps them to survive on one hand and enhance the economic condition on the other. Therefore, there is a link between women and development activities in Karbi society.

Objectives

The study aim to focus the involvement of Karbi women in developmental activities, i.e., the involvement of Karbi women in agricultural activities raise their economic potentiality, i.e., the economic condition of the family such as income standard of living etc., but as they belong to a particular ethnic group which provides certain special customs for the womenfolk. In Karbi society women have to take complete responsibility in the area of income generation.

The following customary practices of ethnic group have to play a very significant role in the economic condition of their family. These customary practices also compel them to go through some vested realities as per expectations of the Karbi society to some activities both outside and inside of the family. Therefore, the observation of customary practices, with regard to development activities, is very significant for the women who involve in development activities. In Karbi society, women are mostly involved in agricultural operation and marketing of their agricultural products. Keeping their customary practices and development activities in mind, following objectives are formulated for the study:

(i) To observe the customary practices of Karbi women in the context of involvement in developmental activities, and
(ii) To study the extent of involvement of the respondents in agricultural activities.

Field and Methodology of the Study

North-East region of the Indian subcontinent includes eight States, i.e., Arunachal Pradesh, Assam, Manipur, Meghalaya, Nagaland, Sikkim and Tripura which are predominantly a region of tribal people. The region is mainly covered by both hilly terrains and plain areas. The entire region constitutes areas of 2,55,083 sq. km of Indian territory having a population of 42.04 million. As per 2010 Census significantly the North-East has more than 2000 km of India's international border with neighbouring countries like China, Myanmar, Bhutan and Bangladesh. The region is considered as a melting pot of different ethnic groups.

As the domain of the study is concerned with Karbi women and developmental activities Lumbajong development Block of the Karbi

Anglong distirct has been selected purposively as the field. Karbi Anglong district comprises of three sub-divisions namely Diphu, Hamren and Bakajan. The Lumbajong development Block belongs to Diphu sub-division. Literally 'Lum' means hill peak and 'Bajong' means 'little ahead'. So the word 'Lumbajong' indicates that reach the hill peak a little ahead. Significantly, the Karbi villages are named after the name of the village *Gaonbudha* (headman) with certain exceptions. Lumbajong development Block includes for villages, *i.e.,* Bura Teran, Dhonsing, Engleng, Harchi Ali and Hemari Timung. The name of the villages could be changed with the death of the headmen subject to the approval of the district council. But at present, the Autonomous council appoints a Governing Body to look after matters of government, besides the local *Gaonbudha.* This system has been continuous since the time of colonial rule.

The Bura Teron village is situated at a distance of 25 km from the district headquaters, Diphu. When Dhonsing Engleng village is away from a distance of 22 km from the same. On the other hand Hemari Timung is located at the distance of 21 km from the district headquarters. Again, Harchi Ali village is located at a distance of 28 km from the district headquarters. The Bura Terang village is bounded by Singhason hills in the west, Borajan in the east, Khonbamon hills in the north and Langparpon river in the south. The total area is about 2 km in breadth and 4 km in length. This village was founded by Mr. Bura Teron in 1960. There is a primary school, namely Siloni L.P. school which was established in 1964. Dhonsing Engleng village is delimited by Khonabamon hills in the North; Bini Taro village in the east; a forest area in the west and the National Highway No. 36 in the south, Mr. Kania Engleng was the founder head of this village. The village has two schools, which started in 1965, one is Borjan L.P school and another is Rangsina M.E. school.

The Hemari Hemai Village is sloppy, hilly and rugged. The paths and the roads take zigzag course. The village is surrounded by upper Mohenjijua area in the east, Binitaro village in the west, Khonbamon in the north and the National Highway No. 36 in the south. The fourth village of the Lumbajong development Block is the Harachi Ali is bounded by Borjan in the east, Khonbamon hills in the north, Longnit River in the west, Bura Teran village in the south. The village covers an area of 2½ km. there is also one school in the village. The name of the village is "Harchi Ali Lower Primary School" which was established in 1984.

The major source of livelihood is agriculture in all the four villages. Besides agriculture they involve in livestock. Major crops grown in the villages are paddy, yam, colocassia, sesame, ginger, chilly and turmaric, some of villagers cultivate 'Muruli' bamboo which provides a good source of income. The four villages have both nuclear and joint families. All the households in the villages are *kutcha.* The houses are neither big nor small. The common building materials are bamboo, thatch, timber, wood, mud and metal wires; electrification in the villages is not satisfactory. Only a very few households have got electricity connection and rest depends on 'hurricane', i.e., kerosene lamp and 'Saki' or earthen lamps.

The following Table 16.1 indicates the population and household of the villages of the Lumbazong development Block.

Table 16.1: Distribution of Population and Households in Lumbazong Block

Sl. No.	*Name of the Village*	*Total population*		*No. of Households*
		Male	*female*	
1.	Bura Teron	215	205	70
2.	Dhonsing Engleng	251	220	71
3.	Harchi Ali	110	126	32
4.	Hemari Timung	109	119	27
	Total	**685**	**660**	**200**

Out of the total female population only 326 are married women. Again out of 326 married women 276 married women as respondents but significantly the number consist of 10 per cent of married women belonging to the age group of 18 to 50. As women in Karbi society are most hard and arduous labourer in agricultural field, it is noticed that after 50 years no women is free from serious waist pain. They suffer from waist pain and feel uncomfortable for their most strenuous posture. Even most of them are struggling to survive physically because of their health conditions. Keeping in the view the research problem, it is decided to carry out the study among those Karbi married women who belong to age group only 18 to 50. Hence, 136 Karbi married women are considered as representative sample of the respondents in the study and the data of the study have been collected from both primary and secondary sources. Interview schedule was prepared for collection of primary data.

SOCIO-ECONOMIC BACKGROUND OF THE KARBI WOMEN

The role of Karbi women in socio-economic aspects is very significant. It is equally important as that of men. But they have no right on property. Even in certain areas women have been considered as second category people of their society. In some social functions and ceremonies they are not allowed to participate like men. But Karbi women are better than many of the women counterparts of various communities in India as they have the right to choose their partner, and have no dowry system, no '*purdah*' system and enjoy freedom of remarriage and certain exclusive privileges to present their cultural skill. Dutta (1993) also opined that women in Karbi society enjoy more or less equal status with that of the men, except in certain occasions. Further he commented that Karbi women are silent spectator than a 'Vocal abitor'

They are not more educated. Till the day of field survey, in this study, it was noticed that only 12 (8.69%) out of 138 married women achieved formal education. Again out of 12 women only 4 (33.3%) studied up to the primary level. The educational level of the married women of the villages under the Lumbajong development Block is very poor and it needs special effort to motivate the people, especially women for pursuing formal education.

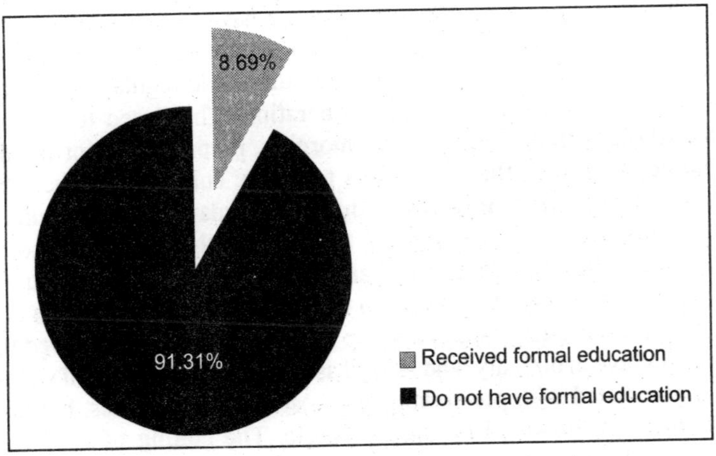

Fig. 16.1: Educational Qualifications of the respondents

In this study half of the women, (69) are in the age group of 30-40 years. Majority of the women, i.e., 112 (81.15% are having nuclear family. Though the earlier Karbi society belonged to joint family mostly, but the vast change has been observed where nuclear family has outnumbered the joint family system. Agriculture is the prime source of income for all women in this study. Out of 138 women majority (86%) of the women have monthly family income in between Rs 2,500 to 3,500. On the other hand this study also reveals that all families are having at least 5 bighas of land for cultivation. It is also observed that out of 138 women, 129 (93.47%) have *kutcha house* and rest of them have semi-pucca house. The housing condition is not in a satisfactory manner. Here, in the study it is found that all those married women who considered as respondents belong to Hindu religion. None of the women belong to any other religion. Majority of the women, i.e., 79 (57.24%) belongs to large family which consists of more than 6 members.

The findings of the socio-economic profile of the Karbi women in this study have given an insight into the work participation in developmental activities. As they are not economically well and good, they have been compelled to involve in agricultural activities. Each family has a considerable area of land for cultivation and involve seriously in agricultural operation to enhance their family income.

Karbi Women and Involvement in Agricultural Activities

In this study, it is found that the women have to remain busy from morning to evening inside home during agricultural season. All women in the four villages visit agricultural field regularly and make themselves busy in agricultural operation. They used to do their household activities early in the morning, prepare food for the day and devote rest of the work hours to '*jhum*' cultivation. When they were asked about their busy schedule in everyday life for agricultural activities they never hesitated to disclose their poor economic condition. They said that doing hard and arduous labour in agriculture is strictly necessary to survive in the society. They opined that only agricultural products are to hike up their economy. It is observed that they involve in both dry land as well as wetland cultivation. All women are associated with pre-sowing and post-sowing activities, but none of them are involved in cutting jungle. The cutting of jungle are commonly fixed for men. Again they also involve in both harvest and post-harvest activities. They use to do the activities like *threshing, winnowing, drying of grains, storage, dehusking, cleaning* etc.

Interestingly; out of 138 women 99 (71.73%) women have the habit of carrying crops from agricultural field to their houses. Of course, it is also found in this study that a few of the women more or less involved in fruit and horticulture; especially pineapple and ginger. They also engage themselves in *Harif* which is called *Bam Kheti* in Assamese. The products of *Bam Kheti* are used to sell in the local market for their livelihood. It is noticed that all the women work 6 to 8 hours a day on an average in the agricultural field. During non-agricultural seasons they work more inside the house. Hence, it is clearly noticed that women spent major time and energy in economic activities which indicates involvement in developmental activities.

Karbi Women and Other Developmental Activities

Apart from agricultural activities Karbi women involve in other allied development activities. In this study it is found that the women associated with weaving, livestock, making bamboo and timber materials, etc. Out of 138 women, 136 (98.55%) women are involved in weaving. They have a set of loom and produce some unique cloths here it is observed that they involve in the activities like spinning of

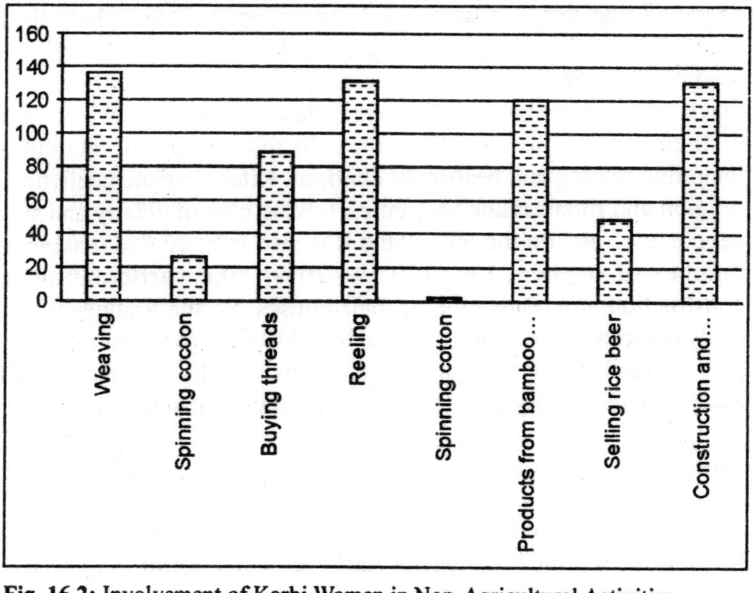

Fig. 16.2: Involvement of Karbi Women in Non-Agricultural Activities

cocoon, spinning of cotton, buying of threads and reeling. But out of the total women in this study, only 26 (18.8%) perform the activity of spinning cocoon. On the other hand 89 (64.49%) showed their interest in buying threads, when 131 (94.92%) women are involved in reeling. Interestingly only 2 (1.44%) women involved in spinning of cotton. In the context of all the activities they see the privilege of marketing for development of their economic condition (Fig. 16.2).

In this study it is also found that a good number of women have been engaged themselves in some of bamboo and timber materials. Out of 138 women 121 (87.68%) women make material out of them 56 (46.28%) women reported that they make materials for both sale and household use. On the whole women in this study involve in most of the activities related to care of animals and birds, specially collection of fodder, cleaning rearing area, preparation of food, etc. Again the study reveals that all women contribute to their family in financial context. They earn money by selling woven clothes, forest products, livestock, bamboo and timber products and horticultural products. In this study one interesting finding has come out that all women are found to engage in preparing a local rice beer which is commonly known as *harlong.* Even out of the total women 48 (34.78%) women earn money by selling this rice beer. Therefore it is observed that they work hard for generating income to help their family. Throughout the time they earn money by selling different kind of products based on the season. Therefore, the study reveals that Karbi women significantly contribute to achieve the development of their family. Besides these they do more domestic activities including rearing and bearing of children. Kitchen related activities, religion and rituals related activities, talking care of elderly and sick people, etc., and on the other hand it is also focused that maximum number of women, i.e., 130 (94.20%) take participation in construction and renovation of their houses. So the findings of the study depict a very clear picture that Karbi women participate directly and indirectly in developmental activities. In this way the role of Karbi women is very significant in developmental activities.

Conclusion

From the above findings it is evident that the active part played by Karbi women is the backbone of Karbi economy. Karbi women have been participating in agricultural activities since inception of this tribal group. This participation is the driving force behind the development of the group and sustains the developmental process. However, this contribution is bounded by tradition as the Karbis are a matrilineal tribe. Agricultural activities were traditionally a responsibility of the Karbi women which suited their matrilineal social structure.

But with the changing nature of global and national economy and with the upgradation of technology, the economic opportunities and working conditions for all sectors have undergone various changes. Now people can work under comfortable situations depending on the nature of work. Modern techniques of agricultural cultivation help to extract maximum output at a grand scale with the use of less exhausting labour.

All the above mentioned changes are not properly visible among the Karbi villages. Karbi women still have to work under physically and mentally strenuous conditions and cannot avail modern techniques due to their poor financial conditions. Even modern medical facilities that have been introduced in the North-East fail to reach out to these women who have to bear the ill effects of their strenuous labour. Hence, the role of Karbi women in sustainable development of their economy become more valuable when there is almost no technical facilities available to them.

REFERENCES

Beteille, A. (1983). *Equality and Inequality Theory and Practice.* Delhi: Oxford University Press.

C. Sridevi (1992). Women Education for Employment. *Social Welfare.* Vol. 46 (2).

Dutta, P. (1993) *The Karbi Women of Assam: A Short Profile.* Guwahati: Assam Institute of Research for Tribal and S.C.

Gogoi, P.K. (1978). *The Economy of Karbi Anglong.* Assam : Directorate of Economics and Statistics.

Gupta, V.K. (2008). *Gender Discrimination and Human Rights.* New Delhi: MD Publications Pvt Ltd.

Ingle, P.O., and S. Kohad and B.H. Rahad (1990). *Participation of Tribal Women in Agriculture.* Maharashtra: Maharashtra Jr. of Ext. Edn.

Kalbagh, C. (1991). *Women and Development,* Vol III. New Delhi: Discovery Publishing Health.

Misra, G. and Chandiramani, R. (2007). *Sexuality, Gender and Rights.* New Delhi: Sage Publications India Pvt. Ltd.

Momsen, C. (1991). *Women and Development in the Third World.* London: Routledge.

Pandya, R. (2007). *Women in India Issues, Perspectives and Solutions.* New Delhi: New Century Publications.

Roy, R.; Miri, S. and Goswami, S. (2007). *Northeast India Development, Communalism and Insurgency.* Delhi: Anshah Publishing House.

Saikia, J.P. (2010). *Gender, Themes and Issues.* Guwahati: Naba Bharati Prakashani.

Saikia, R. (2008). *Role of Karbi Women in Family Research Management, A Doctoral Thesis.* Dibrugarh Assam: Dibrugarh University.

17

Handloom Cottage Industry in Manipur: A Boon for Women Empowerment through Self-Employment

P.C. Sikligar

Empowerment could be inbuilt through capacity building, taking care of human resource development factors, efficacy of the project, skills in marketing and purchasing of raw materials, maintenance of accounts, exposure visits, etc. Handloom Cottage Industry promotes entrepreneurship among the women in the State of Manipur where they have got inborn skill for weaving. The present study focuses on the various aspects of handloom cottage industry in terms of problems, importance of special project (under SGSY: Swarna Jayanti Grameen Swarozgar Yojana), and strategy for further development of handloom cottage industry in Manipur. The District Rural Development Agency (DRDA), Technical, Financial and Management Institutions are trying to promote this industry but despite that this study finds that there is a need for good number of non-governmental organizations, women groups, social activist, dedicated leaders, etc., who can be associated in campaign and support of the cottage industry which absorbed a large chunk of women both in rural and urban areas. This study has been carried out in Imphal districts both east and west to understand the problems and scope of the handloom cottage industry in the State where women are working under the shadow of self-help groups. This study covers focus group discussion with the members of different 20 self help groups which are engaged in weaving.

Introduction

Manipur enjoys a distinct place amongst the Handloom zones in India. Handloom industry is the largest cottage industry in the State. This industry has been flourishing since time immemorial. One of the special features of the industry is that women are the only weavers. According to the National Handloom Census Reports (1988) there are about 2.71 lakh looms in Manipur.

It is believed that Chitnu Tamitnu, a goddess, discovered the cotton and she also produced the yarn. When the threads are ready for weaving she arranged the required equipments and constructed the '*sinnaishang*' (workshed). It is also believed that the goddess Panthoibee once saw a spider producing fine threads and making cobwebs and from it she found the idea of weaving and thus started weaving.

Most of the weavers who are famous for their skill and intricate designing from Wangkhei, Bamon Kampu, Kongba, Khongman, Utlou, etc., in respect of fine silk items. The women got inborn skill to prepare all types of fabrics. Tribal shawls include all varieties of fabrics. Tribal shawls with exotic designs and motifs are the products of five hill districts of the State. Fabrics and shawls of Manipur are in great demand in the national and international market. Today, major handloom production activities are undertaken by three Government organizations, namely, Manipur Development Society (MDS), Manipur Handloom and Handicrafts Development Corporation (MHHDC), and Manipur State Handloom Weavers Co-operative Society (MSHWCS).

About the State

According to the last Census (2001), the total population of Manipur State was figured 23,88,634 of which 60 per cent of the population live in the valley region of the State whereas rest of the population live in the hilly areas. The State population is a combination of non-tribal and tribal people. The Constitution of India has earmarked a total of 29 tribal communities in Manipur. The tribal people mostly inhabitant in the hilly areas, and the plains have a dense population of non-tribal people. In Manipur, more than three-fourth (76.11%) population consisting of 18,18,224 live in the rural areas. The population density in Manipur is 107 per square kilometres. The literacy rate of State is above than 69 per cent. The State has got nine districts such as Imphal (East), Imphal (West), Thowbal, Bishnupur, Churachandpur, Tanenglong Ukhrul, Chandel, and Senapati.

Feasibility of Handloom Occupation

Due to lack of industrialization in the State, handlooms industry play a significant role in economy of the State. Handloom weaving is popular in the State in the form of cottage industry under which women are predominantly engaged. The following reasons actively promote handloom industry in the State such as: Women in the State have got inborn skill in weaving, Women in the State are engaged in the weaving, Women in Manipur, are deeply interested to work under the umbrella of self-help groups, and Handloom products such as bed sheets, pillow covers, chadars, shawls, door and window curtain, mosquito curtain, cotton quilt, etc., are demand based products within and outside the State.

Major Barriers of Handloom Industry

Weaving is one of the popular sources of livelihoods in Manipur which is predominantly engaged by women. The following problems against the smooth functioning of handloom cottage industry have been recorded. Poverty among the skilled women make them unable to obtain required material in time. Lack of working capital and proficiency especially in case of maintaining accounts, marketing, update design, handling financial institutions, etc., make them help less, and to fulfil the wide demand of handloom products in the State, country and abroad, women are helpless to use equipments which promote production.

Market Specific Problems

Market depends on season, capacity of purchase, demand, quality of products, bargaining, interest of the people, etc. In between middle men and seller especially retail and wholesalers play a significant role to boost flexibility of products. Women in Manipur face the following problems of marketing: Lack of enough market supportive organizations, members of the handloom products face problems of marketing, members of groups are compel to sell products on cheap rates to fulfil financial crunches of the family, Superfluous of middle men, nultiplicity of market charge, inadequate store facility, lack of transport facility especially in the rural areas, lack of financial support in time, and absence of grading and standardization of products, etc.

Objectives of the Study

- To understand the problems of handloom weavers in terms of capacity-building, financial constraints and marketing support;
- To understand the scope of handloom and cottage industry; and
- To make recommendations for sustainability of handloom industry in the State.

Methodology

This study has been carried out in Imphal districts both east and west where a large number of women are engaged in the handloom cottage industries from times immemorial. A sample of 20 self-help groups was interviewed through focus group discussion. Each self-help group comprises 10-15 members which include president, secretary, treasurer, and the members. The field work of the study was carried out in 2009 supported by the organised questionnaire.

This study concentrates on different aspects like production and marketing, self-reliance, maintenance of accounts, inter-loan facility, source of credit, utilization of incomes, empowerment of women, etc.

Major Findings of the Study

- Women engaged in handloom weaving have got skill from their childhood days but on the other side, they are very poor in maintenance of accounts, records, ledger, marketing, etc. They are accustomed to sell products locally due to lack of much exposure of the market outside of the State. This process would not give them reasonable price of the product. In other words women are not well equipped in the marketing. They sell products on marginal profit through middlemen or to the local shopkeepers;

- Women face problems to grab new design, therefore, timely skill required to update their skills according to demand of market;
- Women from different social background have got skill in handloom weaving but due to non-availability of operational goods like loom, drums, etc., they are bound to work on others' loom on marginal profit;
- Due to financial constraint women are unable to purchase raw materials in time;
- Women do not have common place where they can work together, and store products, and keep raw materials. They work at home in congested environment;
- Women need linkage with the financial institutions to fulfil credit in time and repeatedly;
- In absence of market channels/centres women face problems of marketing, store of product and raw materials, and placing of the records; and
- Due to lack of Website women unable to get latest information of different kinds like design, market strategy, availability of raw material on reasonable rate, etc.

Special Projects under SGSY: A Suitable Approach for Promoting Handloom Cottage Industry

The Swarnajayanti Gram Swarozgar Yojana (SGSY) is an important programme of Government of India which promotes self-employment through various mechanisms like capacity building, linkage with financial institutions and develops marketing skill, etc. that empowers women. The 15 per cent of funds under SGSY is earmarked for Special Projects by the Ministry of Rural Development, Government of India. The objective of special project is to ensure a time bound programme for bringing a specific number of BPL families above the poverty line through self-employment programmes. The projects involve different kinds of strategies to provide long term sustainable self-employment by organizing rural poor, by creating support of infrastructure, technology, marketing, training, etc.

Recommendations

The Handloom Cottage Industry in the State could be promoted through the Special Projects under Sawrnajayanti Grameen Swarozgar Yojna by paying attention on the following aspects stage-wise:

(a) Formation of Groups

- The self-help groups of women between the age group of 18-55 years belonging to BPL families covering different social groups need to form so under the umbrella of self-help groups and they can obtain various skill, and
- In the process of forming groups, priority needs to give those women who work on others loom on marginal profit. As most of their incomes spend on paying the rent. A sizable number of women despite inborn skill unable to attach with the handloom industry.

(b) Training

- All members need to train in developing of new design, maintenance of cash book, ledger, records, documents, marketing, etc which would empower them further, and
- The update technical skill in handloom technique and professional skill in maintenance of accounts and general development can be imparted through the Handloom Technical Institution, and Institute of Cooperative Management which are located at Stage headquarters. During training period women are expected to facilitate with stipend, to and fro bus fare, accommodation, etc.

(c) Linkages with Financial Institutions

- After training, women need to link with the financial institutions such as Manipur Rural Bank, and Imphal Urban Co-operative Bank Limited so they can obtain credit in time and repeatedly to improve production capacity.

(d) Marketing Support

- Marketing is one of the important aspects of whole production circle which could be promoted by opening centres at village,

Block, and district level and further tieup with the State level selling counters, exhibitions, etc.

Action required for Strengthening of Handloom Cottage Industry in the State

- Prior to implementation of the project, the target group need to consult and need to take in confidence for healthy output of the project. The problems and the notion of the target group need to study;
- The special projects under SGSY need to promote through noble social activists, dedicated non-governmental organizations, goal oriented community based organizations, women welfare associations or youth who believe in self-employment;
- The self-help groups need to form by a Committee comprising of District Rural Development Agency (DRDA); Handloom Department, Financial Institutions, and Elected Representatives especially Ward Members/Community Based Organization;
- The funds should be released based on recommendations of the District Nodal Department supported by document, progress, enthusiasm and dedication of the members, etc.;
- The self-help groups should be asked to maintain records, documents, etc., sequencely;
- Monitoring part of the handloom industry should be handed over to nodal department, handloom department, financial institutions jointly which are well wishers for success of the project;
- The market channels on places like Gram Panchayat, Block, District levels will help in marketing, stores of the products, materials, etc.;
- In Manipur, women are entrepreneurs and they are predominantly engaged in handloom cottage industry. In such situation, tie with the special projects will improve quality of life of the people in the State. The self-help groups could be attached with the prestigious non-governmental organizations for healthy output of the projects, guidance, and;
- Similarly, other cottage industries which are popular among

the youth need to cover under the special projects which will absorb unemployment in the State. The self-employment promotes self-sufficiency among the youth.

REFERENCES

Government of India (1999). *National Handloom Census Reports*. New Delhi: Government of India

Government of India (2001). *Census Report 2001*. New Delhi: Government of India

Sikligar, P.C. (2008). Economic Upliftment of Minorities through Micro-Finance in Indian States: An Evaluation Study in Uttar Pradesh, West Bengal and Karnataka. *Indian Journal of Public Administration*, Vol. LV (2).

Sikligar, P.C. (2004). Women's Empowerment under the Wavli System: A Study on Tribal Women in South Gujarat. *Asia- Pacific Journal of Rural Development*, Vol. XIV (2).

18

Promoting Rural Sanitation in Assam: Issues and Policy Measures

Anupam Hazra

Sanitation connotes a comprehensive package of measures relating to sanitary disposal of human excreta, liquid and solid waste disposal, food and home sanitation, personal and environmental hygiene, safe water storage and handling practices. In current context of development, sanitation is regarded to be one of the key indicators that reflect the quality of life of a particular nation. If water is considered to be 'life', then sanitation must be regarded as 'the way of life' and access to such facilities has an impact on the quality of human life and health. A holistic definition of sanitation includes safe drinking water, liquid and solid waste management, environmental cleanliness and personal hygiene. Failing to ensure any one of these can have direct implications on the individual/family/community's health. According to joint UNICEF and WHO (2012) estimates for 2010, 15 per cent of people in the world openly defecate without any toilet or latrine and are suffering from inadequate sanitation facilities; 60 per cent of these live in India.[1]

Lack of Sanitation: An Emerging Threat to the Millennium Commitments

Sanitation is enshrined in the Millennium Development Goals and is a cornerstone of the fight against poverty (Table 18.1). Achieving

Table 18.1: Inadequate Sanitation and its Impact on Achieving Millennium Commitments

Millennium Development Goals	*Impact of Inadequate Sanitation*
Goal 1: Eradicate extreme poverty and hunger	Lack of clean water and adequate sanitation is a major cause of malnutrition and a significant contributor to extreme poverty. Safe, available water and improved sanitation would enable increased food production and create livelihood opportunities by freeing large amounts of time currently used for water carrying and recovering from disease. Adequate basic services such as water and sanitation can also be key factors in determining the level of private investment.
Goal 2: Achieve universal primary education	The time required to collect and transport water over long distances prevents millions of girls from attending school. In addition, water related diseases cost hundreds of millions of school days each year throughout the world.
Goal 3: Promote gender equality and empower women	Women are most often responsible for collecting and transporting water. Lack of access to safe water and private sanitation facilities prevents girls and young women from attending school.
Goal 4: Reduce child mortality	Poor quality water and inadequate sanitation is the second largest direct cause of child mortality and also contributes to many other deaths by stunting growth and contributing to malnutrition.
Goal 5: Improve maternal health	Disease associated with maternal health can be reduced by the provision of clean water and adequate sanitation.
Goal 6: Combat HIV/AIDS, malaria and other diseases	Lack of clean water, adequate sanitation and hygiene exposes people with HIV/AIDS to increased risk of infection. Poor sanitation increases the risk of malaria. Improving water and sanitation would reduce costs of water and hygiene related illnesses and, therefore, making available additional resources to treat other diseases.
Goal 7: Ensure environmental sustainability	Sustainable treatment of human waste is critical to improving environmental management, especially in areas of dense population
Goal 8: Develop a global partnership for development	Water and sanitation are cross-cutting issues. The development of partnerships to address the lack of access to safe water and sanitation will have flow on effects for other development goals.

Source: Based on Hazra (2013).

the sanitation and water targets would catalyse the attainment of all eight MDGs. Relieving women and girls of their water-fetching burden would help to promote gender equality (MDG3: Promote gender equality and empower women) and boost enrolment as well as attendance rates especially of girls in school (MDG 2: Achieve universal primary education). Good sanitation is directly proportional to attendance in schools and this holds true especially for girls. Children suffering from diseases caused as a consequence of lack of water, sanitation and hygiene perform poorly due to less school hours attended and the reduced capacity to learn. In places where there is acute shortage of water supply and sanitation, children have to fetch water from far off and trudge long distances to defecate. The effort strains their energy and the loss of time inclines them away from education. It is evident from the studies that girls who have been educated have improved maternal health (MDG 5: Improve maternal health). A grave consequence of lack of water, sanitation and hygiene is on the health of pregnant women who stop having food to avoid the inconvenience and physical exertion of open defecation. The suppression of the urge to defecate leads to further physical and physiological complications and ill health. This has a negative impact on the health and nutritional status of the child, both before and after birth. Diseases and ill health reinforce each other so prevention of the diseases caused by lack of water, sanitation and hygiene can make the management of other diseases like HIV/AIDS much easier. (MDG 6: Combat HIV/AIDS, malaria and other diseases). Improved access to clean water would reduce diarrhoea and water-borne diseases, and the transition from unimproved to improved sanitation is accompanied by a 30 per cent plus reduction in child mortality (MDG 4: Reduce child mortality) Water and sanitation related diseases remain among the biggest killers, especially of children. Irrigation accounts for 70-90 per cent of water use in the developing world, and links closely with the MDG1 (Goal 1: Eradicate extreme poverty and hunger) target seeking to reduce by half the proportion of people who suffer from hunger. The sanitation and water targets positioning within MDG7 (MDG 7: Ensure environmental sustainability) makes explicit the link with environmental sustainability. Improving sanitation reduces the contamination of the environment, including water and soil, thus protecting biodiversity and people's livelihoods in a sustainable manner. A healthy living environment is closely connected to sanitary toilets, especially in dense settlements. Developing and strengthening international networks for sanitation and water would contribute towards the attainment of MDG8 (MDG 8: Develop a global partnership for development) seeking a global partnership for development. (Hazra, 2011).

As far as the scenario of rural India is concerned, sanitation related diseases take a heavy toll of lives, especially children's lives, and are a drain on productivity and incomes. The global impact of poor sanitation on infant and child death and health is profound. Lack of adequate sanitation also forces households into the continued indignity of open defecation, which is an acute problem especially for women and young girls. Lack of sanitation and poor hygiene severely limit the impact of other development interventions in education, health, rural and urban development. Providing proper sanitation facilities to millions of people is a significant challenge, especially in the world's second most populated country. The task is doubly difficult in a country where the introduction of new technologies can challenge people's traditions and beliefs. A humble clean toilet advances progress towards gender equality, promotes social inclusion, increases school attendance, builds community pride and social cohesion and contributes to poverty reduction. Improving access to sanitation is, therefore, appropriately included in the Millennium Development Goals and if we have to achieve the status of developed nations, we certainly need to address the issue of sanitation more seriously, which is quite low at present especially in the rural parts of our country.

Rural Sanitation Scenario in India

According to the National Family Health Survey 3, the coverage of rural sanitation is 26 per cent. The urban sanitation coverage according to the NFHS-3 is 83.2 per cent. The all-India coverage of sanitation, according to the survey is 44.6 per cent in 2005-06 which is 8.9 per cent increase from 1998-99 survey of NFHS-2. (IIPS and Macro International, 2007). However, studies across rural India have pointed out to low levels of latrine usage because of lack of awareness of the importance of sanitation, water scarcity, poor construction standards and the governments' past emphasis on expensive standardized latrine designs. Initial indications of an evaluation by the Government of India and UNICEF's Child Environment Programme show that significant number of people, especially Below Poverty Line (BPL) households, are not using their latrines. If variables such as usage are also included in a definition of sanitation coverage, the national picture is likely to be worse with a corresponding impact on the number of people who need to be reached to meet the Millennium Development Goals (MDG) target. Low sanitation coverage in rural India is primarily due to insufficient motivation/awareness of people and lack of affordable sanitation technology. People (mostly from lower economic strata) are generally not aware of the health and environment benefits of sanitation and it is still not a "felt needs" for them, resulting in absence of people's participation in sanitation programmes. Non-availability of choice of toilet designs, area specific technologies, inadequate supporting delivery systems and absence of trained masons, skilled workers and technical manpower are also reasons for low coverage.

Lack of adequate sanitation is a pressing challenge in rural India. Everyday, an estimated 1,000 children under five years die in the country because of diarrhoea alone. Prevalence of child under-nutrition in India (47 per cent according to National Family Health Survey (NFHS) III, 2005-06) is among the highest in the world.[2] Child under-nutrition is aggravated by the prevalence of diarrhoeal disease, and is responsible for 22 per cent of the country's burden of disease (Gragnolati, *et al.*, 2005). For one of the most densely populated subcontinents in the world, India has the lowest sanitation coverage. A comprehensive strategy for promoting sanitation in rural areas of Indian States was first developed during the International Water Supply and Sanitation Decade (1980-90). The percentage of rural households with a toilet in their own premises at the beginning of the decade was less than five per cent. The formulation and launching of the Centrally-Sponsored Rural Sanitation Programme (CRSP) was the first policy statement at the national level. The programme provides central assistance to the States to construct standard two-pit design toilets in the premises of households belonging to the socially and economically disadvantaged sections. Affordable sanitation remains a challenge in the situation where approximately 100 million new latrines are needed to increase the coverage from the present 15 per cent to 85 per cent or more. Twenty-four million new latrines were estimated to be built before the year 2001. A drastic increase in construction of latrines has to take place, requiring effective methods for commercial and social marketing and cost effective technology.

State of Rural Sanitation in Assam

Keeping in view the different development parameters, the State of Assam is considered to be one of the very disadvantaged, poor and underdeveloped States in India, like the other North-Eastern States viz., Meghalaya, Manipur, Nagaland, Mizoram, Arunachal Pradesh and Tripura. The 2011 Census of India report on houses, household amenities and assets for Assam, which was released here on Friday, revealed that there was no significant improvement in providing sanitation facilities to people between 2001 and 2011. The census covered 3.11 crore population in 26,395 villages, 214 towns, including 126 census towns. In 2001, 64.6 per cent households had proper sanitation facilities within their premises. In 2011, there was only a slight improvement of 64.9 per cent. The number of households not having such facilities has also declined to 35.1 per cent in 2011 from 35.4 per cent in 2001. Another picture of Assam highlighted in the census report is that 80 per cent of the households still use firewood, crop residue, cow dung cakes and coal for cooking, while only 19 per cent use LPG or PNG, electricity and biogas. While 20 per cent have drainage facility, 79.59 per cent still don't enjoy such facilities against 79.6 per cent in 2001. (Mitra, 2012)

Table 18.2: Physical Progress of Total Sanitation Campaign (TSC) in Rural Assam (as on 24-5-2012)

Components	*Project Objective*	*Project Performance*	*Percentage Achievement*
IHHL* BPL**	2220017	1528924	68.87
IHHL APL***	1161020	416648	35.89
IHHL Total	3381037	1945572	57.54
School Toilet	34772	33722	96.98
Sanitary Complex	211	57	27.01
Anganwadi	16819	10956	65.14
Rural Sanitary Mart	115	61	53.04

Source: Website of TSC [**Individual household latrines.* ***Below Poverty Line.* ****Above Poverty Line*].

The practice of open-defecation is still prevalent in the rural Assam.[3] Different reports, researches and surveys in the context of sanitation across rural Assam aptly indicate that sanitation has never been a priority, both for government and individuals. Statistics shows that in rural areas, only 8.61 per cent households in 2001 have a pour-flush sanitary latrines, connected to a septic tank.[4] In 1992-93, 35.6 per cent rural households had some kind of a pit toilet or improvised latrines, which are neither flushed nor sanitary.[5] The rest of the households go for open-air defecation. This is a continuous source of pollution of any available surface water source nearby. As a result, the villagers suffer from diarrhoea, intestinal and other water borne diseases several times in each year, resulting in poor health, loss of nutrition, loss of working days, loss of income and finally lower life span. Children are the worst affected. Women, who look after the children, are further burdened. In major parts of rural Assam, lack of sanitation has been the root cause of many of the health and hygiene problems.

The rural sanitation programme in Assam started in 23 districts of the State in 1986 under Central Rural Sanitation Programme of Government of India. Till 1993, only 842 household latrines were constructed in the districts under this program. From 1993, an Intensive Sanitation Project, supported by UNICEF, was also taken up in Kamrup district. About 12,000 single-pit/two-pit pour-flush sanitary latrines have been built in the district up to1998 under this programme, through the Public Health Engineering Department (PHED) of the State. Initially the department produced mosaic sanitary pans and RCC concrete slabs for the latrine and sold the same to the rural families at a subsidized price. Later this activity was assigned to NGOs, to establish Sanitary Marts for production of the components, supported financially by UNICEF. This activity somehow continued up to 2003, with very little progress. UNICEF finally withdrew their support due to various reasons. (DBHP 2005) (Table 18.3).

The most important point to be noted in the existing rural settings in Assam is that the villagers do not feel the need for a pour flush sanitary latrine, as they are accustomed to the existing practice of defecation in the open or use of insanitary pit latrines for generations. Extraordinary efforts for advocacy at the household level (and not the usual kind of meetings and training) coupled with financial and technical support at their doorsteps would be necessary to change the mindset of the people. Only then, sanitation and hygiene improvement in all the households in any selected village would be possible. One or two rich families of each village construct a pour flush sanitary latrine with their own money and effort when they wish,

Table 18.3: Different Models of Toilets being used in Rural Assam: At a Glance

1. **Slab Placed over earthen pit without waterseal:** A dug earthen pit over which a RCC slab is put to cover the pit. The slab has a hole at the centre for defecation. There is no provision for escape of foul gas except through the hole and therefore the latrine smells a lot. Contamination is less than that of open pit latrines.
2. **Slab Placed over brick/RCC lined pit without waterseal:** The pit is much deeper and lined with brick or stone. It is covered with a RCC slab. This is a better variant of the "slab placed over earthen pit" model and lasts for many years. However, it is more expensive to dig a deep hole and line the same with bricks.
3. **Raised wooden/bamboo platform and separate pit connected by an open channel made of galvanized iron sheet (there is no waterseal):** A raised platform (about 0.6 metres high) is built with bamboo or timber. A plain galvanised iron sheet channel is fixed from the centre of the platform to end in a pit at a slope. An empty bitumen drum can be placed as lining for the pit, the open end of the drum being placed inside the pit. Water is used for ablution and for flushing the faeces to the pit. In this case, the pit being away from the defecating platform, there is less smell and also less pollution.
4. **Ventilated Improved Pit (VIP) Latrine:** This is a dry latrine. Outside the latrine wall, a vent pipe is provided, facing towards the sun and covered by a net on top. The pipe is painted black on the outside to absorb heat during sunshine. This helps in creating a draft for the foul gas to escape. There is a hole in the slab which is covered by a removable cover when not in use. This type of latrine is suitable in dry areas where water is scarce. Assam has a very humid and wet climate and this type of latrine is not suitable and also not used, except rarely in certain hilly areas.

Source: DBHP (2005).

will not have any impact on the total sanitation improvement of the villages. Unless the whole village is taken up with a time-bound and determined effort for the construction of sanitary latrines in each house, any lackadaisical programme will not be able to control the ever-increasing level of pollution.

Promoting Rural Sanitation in Assam: The Task Ahead

Increasing sanitation coverage in rural areas would require more clarity of the issue and understanding of the rural sensibilities. Building toilets is just one half of the battle; the other half is to make people use them. The number of villages winning *Nirmal Gram*

Puraskar may be inspiring only if uncovered villages are not taken into count. (Hazra, 2010) While talking about promoting rural sanitation in Assam, it may be stated there is none to bell the cat – to actually construct latrines house to house, to provide technical support and to meet the quality, expected by the people within their affordability. Further there is no provision for loan or grant to the families, except for a little subsidy given by the Panchayat (the local self-government), to the families falling below the poverty line only. Interestingly, the BPL families cannot afford to pay even the required subsidized price for the latrine slab at one time and the latrine is not their priority. At best, a little less poor neighbour can purchase a latrine slab using the poor family as a go between. Most importantly, covering only the BPL households will not ensure that all the rural households will be covered (Box 18.1).

BOX 18.1: Factors to be taken into consideration while implementing sanitation programme in Rural Assam

- Scaling up, social acceptance of the "Total sanitation Approach": peer-pressure to promote compliance.
- Understanding of local motivation for sanitation is essential.
- Wide menu of technological, payment and cost options is key.
- Sanitation access and coverage must be defined and monitored.
- Sanitation should be mainstreamed in poverty reduction strategies.
- Subsidies must be targeted to the poorest.
- Promotion must be gender-sensitive and gender-inclusive.
- Necessary community empowerment requires long term participatory processes.
- Sanitation projects should include hygiene promotion.
- Sanitation projects need to develop a local "industry" (masons, social marketers, hygiene educators, etc.) to sustain services.

Source: Based on Hazra (2010).

Low sanitation coverage in rural India is primarily due to insufficient motivation/awareness of people and lack of affordable sanitation technology. People (mostly from lower economic strata) are generally not aware of the health and environment benefits of sanitation and it is still not a "felt needs" for them, resulting in absence of people's participation in sanitation programmes. Non-availability of choice of toilet designs, area specific technologies, inadequate supporting delivery systems and absence of trained masons, skilled workers and technical manpower are also reasons for low coverage.

(Hazra, 2011) Thus it is obvious that the government and the Panchayats (local self-government) alone cannot make speedy progress in sanitation promotion in rural areas. We will have to wait for half a century to find that not even 50 per cent of the rural families have sanitary latrines. Hence, there is an urgent need for the private sector to step in for sanitation promotion and the need for giving loan to families constructing sanitary latrines. Improvement of sanitation in rural areas is not a priority consideration of the government. Therefore, NGOs and private agencies need to make an effort in this sector. Keeping in view the dynamics of rural Assam as well as the experiences gained from the earlier rural sanitation initiatives, the following measures may be effective in intensifying the efforts towards promoting rural sanitation (Box 18.2):

The sanitation programmes should not be confined to construction of latrines only. Awareness of sanitation standards and health impact of unsanitary conditions must be spread widely. It should be promoted as a total package consisting of safe handling of drinking water, scientific disposal of waste water, safe disposal of human excreta including child excreta, solid waste management, domestic sanitation and food hygiene, personal hygiene and community hygiene. At the same time, there is a need to:

- Streamline the process and to make sanitation a profitable business.
- Market low-cost sanitation like any other commercial product.
- Create "need" for sanitary latrines.
- Most importantly, produce different models.
- Continue Research & Development work for making improvements and provide choices to the villagers to choose a model they like, at the lowest possible price.
- Provide easy instalments. Let the people decide what they want.

Simultaneously, various Non Governmental Organisations (NGOs), Civil Society Organisations (CSOs), women Self-Help-Groups (SHGs),Voluntary Organisations (V.Os) and other Community-Based Organisations (CBOs) of rural Assam should properly be mobilised to support the sanitation programme of the State, especially for —

Box 18.2: Promoting Rural Sanitation in Assam: Some Suggestive Measures

- Encouraging more and more fundamental research in the sphere of rural sanitation with a special focus on different issue of environmental sanitation of rural India.
- Encouraging the development of indigenous, affordable, eco-friendly and culturally acceptable sanitation technologies in rural India.
- Greater emphasis on personal hygiene and environmental sanitation as essential prerequisites for achieving total sanitation status in rural villages.
- Generating awareness and creating demand at community level for safe access to water and sanitation through community participation.
- Rural women are mostly affected by the absence of sanitary latrines at home. Access to adequate and sanitary latrines is a matter of security, privacy and human dignity, particularly for rural women. So while carrying out total sanitation campaign in rural India—targeting gender issues in communication strategy will be quite effective in focussing on the needs of the women and yields considerable success.
- Higher degree of transparency in programme-implementation leading to the proactive provision of relevant and complete information to all stakeholders to bring about their more effective participation.
- Creating stakeholders' network comprising eminent personalities, civic bodies, political elected representatives, industry, educational institutions, civil society, media and community leaders to ensure smooth functioning and sustainability of rural sanitation programmes.
- Environmental regeneration and improvement of environmental resource-base as a source of labour-intensive growth, while augmenting productive capacities, increasing resource-use efficiency and correcting regional and rural-urban imbalances,
- Integration of poverty reduction and environmental regeneration through participatory watershed management.
- Promoting behavioural change towards adoption of environment-friendly practices among the rural mass.
- Creating provision for livelihood security to resource-poor households through sustainable access to basic needs such as food, fodder, fuel and water.
- Initiating location-specific interventions in promoting rural sanitation, keeping in view the diverse natural resource and socio-economic conditions of the nation.

Source: Based on Hazra (2012).

- Supervision.
- Monitoring.
- Organising need-based training programmes.
- Generating awareness about the importance of sanitary practices at the grass-root level/community level/village level.

- Developing innovative IEC (information education communication) materials for the campaign.
- Framing culture-bound approach and strategy (use of rural folk, etc.) for promotion of the programme.

A suitable provision for the participation of these organisations in the sanitation should be made in the project costs.

Promotion of sanitation and hygiene in rural societies of Assam require high level expert plans and a lot of face-to-face advocacy at the grass roots. The planners themselves need to be highly qualified and at the same time need to have enough grass roots experience to make viable projects. Grass roots level personnel have to be suitably trained and their activities monitored. The cost of advocacy and marketing is almost as much as the cost of the latrine; while studies have explored that most of the programmes of the Government does not budget for advocacy and marketing. Last but not the least, the rural sanitation policies of Assam will have to ensure availability of affordable as well as culturally-accepted sanitary technologies especially when rural population are being addressed.

NOTES

1. WHO/UNICEF (2012).
2. IIPS and Macro International (2007).
3. DBHP (2005).
4. *bid.*
5. *ibid.*

REFERENCES

Binoy's Hygiene Promotion Private Limited (DBHP) (2005). *Making Rural Sanitation & Hygiene Viable: A Project Report and Proposal.* India: Ruralsan, 2005. On-line.

Gragnolati, Michele, *et al. India's Undernourished Children: A Call for Reform and Action, Health, Nutrition and Population (HNP) discussion paper*. Washington DC: World Bank, 2005. Print.

Hazra, Anupam. "Rural Sanitation: Still A Mmajor Concern." *South Asia Politics.* 8.10 (2010): 35-39. Print.

Hazra, Anupam. "The 'Sanitation' Crisis in India: A Threat to the 'MDG-

Commitments'." *Perspectives in Social Work* XXVI.1 (2011): 18-23. Print.

Hazra, Anupam. "Rural Sanitation and the Need to Encourage Women's Participation:" *Kurukshetra,* 61.3 (2013): 8-10. Print.

Hazra, Anupam. "Promoting Rural Sanitation for a Better Environment in Rural India." *Kurukshetra,* 60.8 (2012): 11-13. Print.

International Institute for Population Sciences (IIPS) and Macro International. *National Family Health Survey (NFHS-3), 2005-06.* Mumbai: IIPS, 2007. Print.

Mitra, Naresh. "Sanitation still a far cry in progressive Assam." *The Times of India.* 30 March 2012. Print.

TSC (Total Sanitation Campaign). "Physical Progress Report". *Total Sanitation Campaign India,* May 24. Web. 25 May 2012. On-line

WHO/UNICEF. *Progress on Drinking Water and Sanitation: 2012 Update.* Geneva: WHO/UNICEF Joint Monitoring Programme Report for Water Supply and Sanitation, 2012. On-line.

Index